INTERNATIONAL SERIES OF MONOGRAPHS IN PURE AND APPLIED BIOLOGY

Division: **ZOOLOGY**

GENERAL EDITOR: G. A. KERKUT

VOLUME 48

THE PILCHARD

BIOLOGY AND EXPLOITATION

OTHER TITLES IN THE ZOOLOGY DIVISION

General Editor: G. A. KERKUT

Vol. 1. RAVEN—*An Outline of Developmental Physiology*
Vol. 2. RAVEN—*Morphogenesis: The Analysis of Molluscan Development*
Vol. 3. SAVORY—*Instinctive Living*
Vol. 4. KERKUT—*Implications of Evolution*
Vol. 5. TARTAR—*The Biology of Stentor*
Vol. 6. JENKIN—*Animal Hormones—A Comparative Survey. Part 1. Kinetic and Metabolic Hormones*
Vol. 7. CORLIS—*The Ciliated Protozoa*
Vol. 8. GEORGE—*The Brain as a Computer*
Vol. 9. ARTHUR—*Ticks and Disease*
Vol. 10. RAVEN—*Oogenesis*
Vol. 11. MANN—*Leeches (Hirudinea)*
Vol. 12. SLEIGH—*The Biology of Cilia and Flagella*
Vol. 13. PITELKA—*Electron-microscopic Structure of Protozoa*
Vol. 14. FINGERMAN—*The Control of Chromatophores*
Vol. 15. LAVERACK—*The Physiology of Earthworms*
Vol. 16. HADZI—*The Evolution of the Metazoa*
Vol. 17. CLEMENTS—*The Physiology of Mosquitoes*
Vol. 18. RAYMONT—*Plankton and Productivity in the Oceans*
Vol. 19. POTTS and PARRY—*Osmotic and Ionic Regulation in Animals*
Vol. 20. GLASGOW—*The Distribution and Abundance of Tsetse*
Vol. 21. PANTELOURIS—*The Common Liver Fluke*
Vol. 22. VANDEL—*Biospeleology—The Biology of Cavernicolous Animals*
Vol. 23. MUNDAY—*Studies in Comparative Biochemistry*
Vol. 24. ROBINSON—*Genetics of the Norway Rat*
Vol. 25. NEEDHAM—*The Uniqueness of Biological Materials*
Vol. 26. BACCI—*Sex Determination*
Vol. 27. JØRGENSEN—*Biology of Suspension Feeding*
Vol. 28. GABE—*Neurosecretion*
Vol. 29. APTER—*Cybernetics and Development*
Vol. 30. SHAROV—*Basic Arthropodan Stock*
Vol. 31. BENNETT—*The Aetiology of Compressed Air Intoxication and Inert Gas Narcosis*
Vol. 32. PANTELOURIS—*Introduction to Animal Physiology and Physiological Genetics*
Vol. 33. HAHN and KOLDOVSKY—*Utilization of Nutrients during Postnatal Development*
Vol. 34. TROSHIN—*The Cell and Environmental Temperature*
Vol. 35. DUNCAN—*The Molecular Properties and Evolution of Excitable Cells*
Vol. 36. JOHNSTON and ROOTS—*Nerve Membranes. A Study of the Biological and Chemical Aspects of Neuron-glia Relationships*
Vol. 37. THREADGOLD—*The Ultrastructure of the Animal Cell*
Vol. 38. GRIFFITHS—*Echidnas*
Vol. 39. RYBAK—*Principles of Zoophysiology—Vol. 1*
Vol. 40. PURCHON—*The Biology of the Mollusca*
Vol. 41. MAUPIN—*Blood Platelets in Man and Animals Vols. 1–2*
Vol. 42. BRØNDSTED—*Planarian Regeneration*
Vol. 43. PHILLIS—*The Pharmacology of Synapses*
Vol. 44. ENGELMANN—*The Physiology of Insect Reproduction*
Vol. 45. ROBINSON—*Genetics for Cat Breeders*
Vol. 46. ROBINSON—*Lepidoptera Genetics*
Vol. 47. JENKIN—*Control of Growth and Metamorphosis*

The Pilchard by Jonathon Couch, F.L.S., 1869.

THE PILCHARD
BIOLOGY AND EXPLOITATION

MICHAEL CULLEY
Department of Biological Sciences
Portsmouth Polytechnic

PERGAMON PRESS
OXFORD · NEW YORK · TORONTO
SYDNEY · BRAUNSCHWEIG

Pergamon Press Ltd., Headington Hill Hall, Oxford
Pergamon Press Inc., Maxwell House, Fairview Park, Elmsford, New York 10523
Pergamon of Canada Ltd., 207 Queen's Quay West, Toronto 1
Pergamon Press (Aust.) Pty. Ltd., 19a Boundary Street, Rushcutters Bay, N.S.W. 2011, Australia
Vieweg & Sohn GmbH, Burgplatz 1, Braunschweig

First edition 1971

Library of Congress Catalog Card No. 73-144138

Printed in Great Britain by A. Wheaton & Co., Exeter

08 016523 0

CONTENTS

PART IV. THE PILCHARD INDUSTRY OF SOUTH AND SOUTH WEST AFRICA

PART V. STATISTICAL APPENDIX

PREFACE AND ACKNOWLEDGEMENTS

WITH the world population being expected to double in the next 30 years and with the world production of fish having already more than doubled since the war, the need to apply biological knowledge to marine fishery resources becomes rapidly more urgent. Although this book is dealing with only one small aspect of applied biology it is hoped that it will add emphasis to those warnings, already reiterated many times, that marine fish stocks are not inexhaustible, that they need careful management and, an ever-increasing problem this, that international co-operation is required if their management is going to be successful. I hope that the reader will find that the sardine and pilchard fisheries dealt with in the book illustrate these points quite vividly.

Part I of the book consists of a general introduction to the study of fisheries, a biological introduction to the species to be studied in more detail in Parts II, III and IV, and a chapter giving some background material concerned with production in the sea. Readers familiar with this matter will find that it is quite possible to start the book at Chapter 4. Part II, dealing with the English pilchard and its industry, is based on a Ph.D. thesis on the subject and is rather more detailed than Parts III and IV. These two parts are concerned with the California sardine industry and the South African and South West African pilchard industry, respectively. Finally, in Part V are some tables showing the landings of sardines or pilchards of the other countries in the world which catch more than 10,000 metric tons per annum. Where possible, tables showing the utilization of these catches are also shown.

I have become indebted to many people during the course of writing this book. First of all to Portsmouth College of Technology (now Portsmouth Polytechnic) for a research post enabling me to study the Cornish pilchard industry. This post was in the Marine Resources Research Unit of the College and I am grateful to the Director, Dr. R. L. Morgan, for suggesting and outlining the course of study. During that time also I received valuable help from Mr. J. P. Bridger of the Fisheries Laboratory, Lowestoft, Mr. A. Sharples of the White Fish Authority at Plymouth and many other people, especially the fishermen and others directly connected with the industry.

The second stage, that of expanding my thesis into a book, was suggested by the late Professor G. E. Newell, and made possible by a further research post at Portsmouth College of Technology. I am grateful to numerous people for their help during this work and I cannot mention them all individually; however, to the following go my special thanks: Miss Frances Lanaway of the College Library, Dr. Frances N. Clark, formerly of the California Division of Fish and Game, who read the manuscript of Part III and kindly gave me much help with it, Dr. A. J. Southward of the Marine Laboratory, Plymouth, to whom I am particularly indebted for reading the whole of the manuscript and offering much constructive criticism and advice, and finally, to my wife who has helped in many ways throughout the production of the book.

Acknowledgements to sources are given at relevant points in the book, but I must thank the following for permission to use particular illustrations: Dr. N. B. Marshall of the British Museum (Natural History) for Fig. 2.2, Dr. D. P. Wilson of the Marine Laboratory, Plymouth, for Fig. 4.15, plates 1 to 5, Routledge and Kegan Paul for Figs. 7.6 and 7.7 and Crown Copyright for Figs. 5.1, 5.3, 5.5, 5.6, 5.7, 6.2, 7.3, 7.4, and 15.1. A list of figure sources and acknowledgements are given on p. ix. I should also like to thank Irvin and Johnson Limited for a copy of their book *South African Fish and Fishing*, the Fish and Wildlife Service of the United States and the California Division of Fish and Game for many complimentary copies of research reports which they have sent to me.

MICHAEL CULLEY

SOURCE MATERIAL AND ACKNOWLEDGEMENTS FOR FIGURES

Figure	*Source*	*Acknowledgements*
Frontispiece	Couch, Jonathon (1869)	Living descendants of Dr. Couch
1.1	F.A.O. *Yearbooks of Fishery Statistics*	—
2.2	Marshall, N. B. (1965)	Dr. N. B. Marshall
4.3, 4.9, 4.10	Hickling, C. F. (1945)	—
4.11–4.14	After Cushing, D. H. (1957)	—
4.15 plates 1–5	—	Dr. D. P. Wilson
4.16	Cunningham, J. T. (1891)	—
5.1, 5.3, 5.5, 5.6	Cushing, D. H. (1957)	Crown Copyright
5.2, 5.4	—	Mr. J. P. Bridger
5.7	Cushing, D. H. (1957)	Crown Copyright
5.8	After Lumby, J. R. (1925)	—
5.9	After Russell, F. S. (1939)	—
5.10	After Cushing, D. H. (1957)	—
6.1	Information from Bennett, W. J. (1952)	—
6.2	Davis, F. M. (1958)	Crown Copyright
7.1	Great Britain (1946–1967)	—
7.2	Great Britain (1948–1966)	—
7.3, 7.4	Cushing, D. H. (1957)	Crown Copyright
7.5	Hjort, J. (1926)	—
7.6, 7.7	Hodgson, W. C. (1957)	Routledge & Kegan Paul
7.8	Buys, M. E. L. (1959)	—
7.9	Great Britain (Min. Ag. Fish. Fd. monthly statistics)	—
8.3, 8.4	Hickling, C. F. (1939)	—
8.6	After Hodgson, W. C. and Richardson, I. D. (1949)	—
10.1–10.4	After Hickling, C. F. (1945)	—
11.2	Cushing, D. H. (1957)	Crown Copyright
12.4–12.6	Great Britain (1957–1961)	—
13.1	Great Britain (1962a)	—
13.2	Great Britain (1966a)	—
13.3	Great Britain (1962)	—
13.4	Great Britain (1966)	—
15.1	Graham, Michael (1956). Based on a Crown Copyright photograph and reproduced with permission of The Controller, H.M. Stationery Office	Crown Copyright and The Controller, H.M. Stationery Office
16.1	Clark, F. N. (1952)	—
16.2, 16.3	Scofield, E. C. (1934)	—
17.1	Marr, J. C. (1960a)	—
17.2	Radovich, John (1962)	—
20.1	Davies, D. H. (1957)	—
20.2	Davies, D. H. (1958a)	—
21.1	Davies, D. H. (1958b)	—
22.1–22.6	Information from F.A.O. *Yearbooks of Fishery Statistics*	—

PART I

INTRODUCTION

CHAPTER 1

THE IMPORTANCE OF PELAGIC FISHERIES THE SCIENCE OF THE STUDY OF FISHERIES

FISH can be divided into two main groups according to whether they live mostly on or near the sea bed (demersal) or are found in the waters above (pelagic). This division is important to fisheries because different techniques are required for the capture of the two groups. Generally speaking conventional trawls and long-lining are used for catching demersal fish, and drift nets, encircling nets and mid-water trawls are used for the pelagic fish, and very often the boats employed are totally different in character.

The major portion of the world catch of fish by weight is formed by the pelagic group which includes herrings, sardines, pilchards and mackerel as well as larger fish such as tunas and marlins. The most important of the demersal fish are those belonging to the cod family and the haddocks and hakes. The graph in Fig. 1.1 shows how the pelagic fraction of the world catch of fish as represented by herrings, sardines and anchovies, increased over the 12-year period from 1953 to 1964; the catches of tunas and mackerels are shown separately. The major part of this increase has come from the anchovy catch of Peru which increased from 147,800 metric tons in 1953 to 8,863,000 metric tons in 1964.† Pilchard catches in South and South West Africa also soared from practically nothing just after the Second World War to 256,100 tons in 1964 for South Africa and 655,900 tons for South West Africa. (It will be shown later that the more recent figures for South Africa have declined since the 1964 level.) Most of the catches from these countries have recently been reduced to fish meal. (Food and Agricultural Organization of the United Nations—henceforth referred to as F.A.O.—1965.)

From these figures it can be seen that the maintenance of the pelagic fisheries is of great importance if the needs of the world, especially of the emergent countries, for supplies of protein are to be fulfilled. In trying to meet these nutritional requirements it is essential that the fish stocks are husbanded to the best of man's ability, which is circumscribed, of course, by the extent of his knowledge and technical achievement. One way in which this aim can be reached is by an all-out attempt to insure that the stocks are not over-fished. Although this point has been stated so many times that it has become a truism, the self-discipline of the fishing nations of the world is apparently still not great enough to impose catch or mesh limits to aid conservation; or if they are imposed, the countries concerned find it difficult to adhere to, or enforce them. To quote an example from the marine mammals, the blue whale was in danger of extinction in 1962–3 in spite of early efforts at conservation made by the International Whaling Commission in 1954, 1955 and 1960. In 1963 the situation had become so bad that catching of the blue whale had to be completely banned (Mackintosh, 1965). Other examples from stocks thought to be inexhaustible can also be mentioned; the cod grounds off the Lofoten Islands are showing marked symptoms of having been too heavily

†All quantities expressed in tons refer to the metric tons unless otherwise stated.

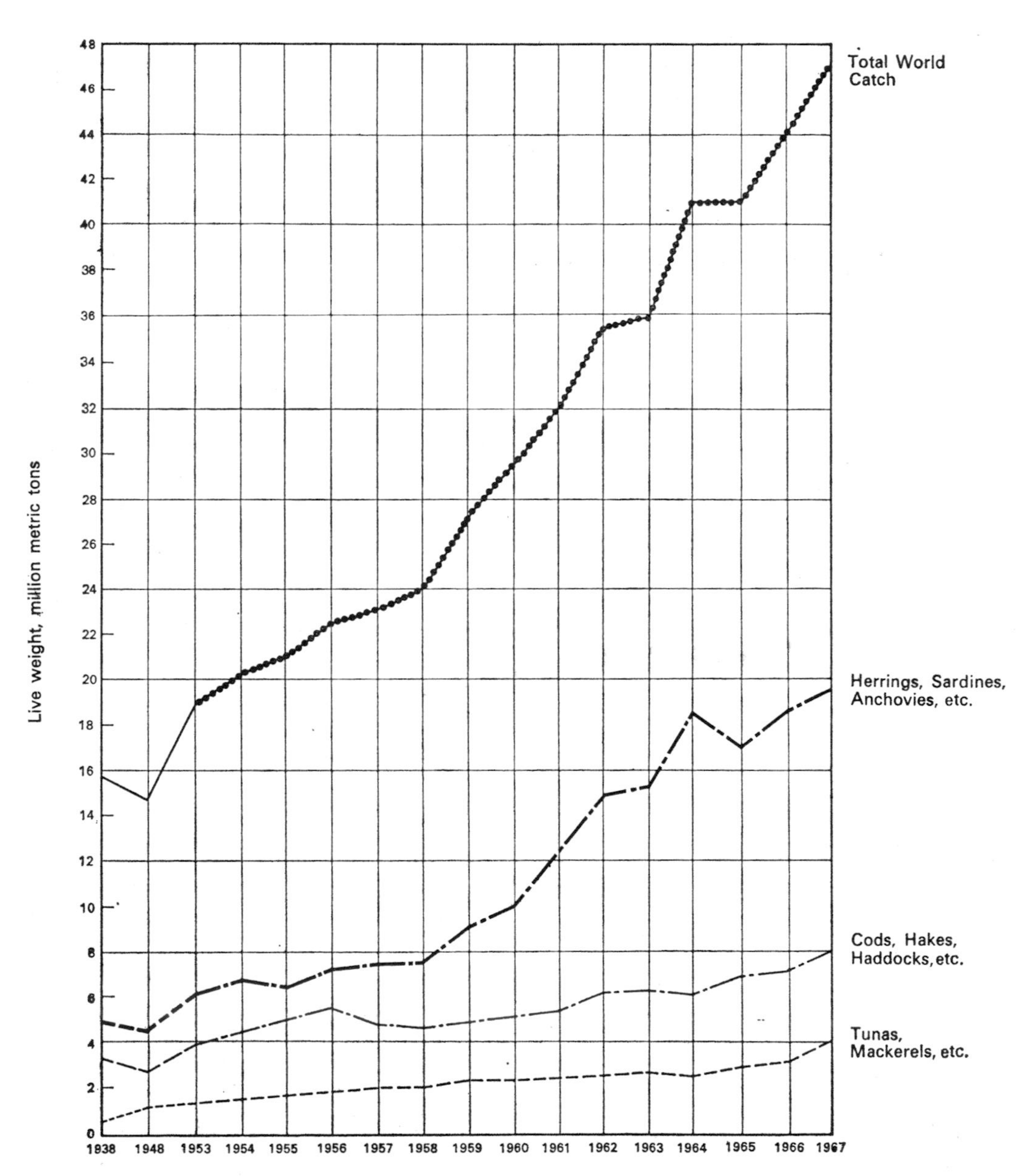

FIG. 1.1. World catch of fish showing the importance of the pelagic fraction.

fished in spite of the great knowledge of fisheries of the two countries most intimately concerned which are Norway and the Soviet Union; the herring stocks of the North Sea are so depleted that the formerly flourishing herring industry of the east coast of England is now relatively depressed. These examples illustrate that the nature of man's predations in the fisheries field is short-sighted and, apart from being biologically tragic, leads to waste in terms of the investment that has occurred in the vessels and gear, since such investment will fail to bring in reasonable dividends when used on a run-down stock.

Fisheries management, or careful husbandry of stocks, will be a recurring theme throughout this book. It is particularly important in the two largest industries considered, those of California and South and South West Africa, and it is hoped that the complex and intriguing nature of some of the problems involved will be shown. An outline of some of the obstacles to be overcome and the disciplines involved in fisheries studies is given in the next few pages.

In the study of a fishery, or any aspect of marine life, immediate difficulties are imposed by the nature of the environment. There are many variable factors to be taken into account and, whereas in terrestrial environmental studies, it is often possible to stabilize one or more of these, in the marine environment this is virtually impossible. Currents, temperature, salinity, the effects of wind, evaporation and precipitation, concentration of nutrient salts, quantity of insolation are among the many physical or abiotic variables acting upon the marine environment. Alteration of any one of these can bring about changes in some of the others and the action of each one has some effect on the biotic components of the environment. The biotic and abiotic factors having an effect on "the area occupied by the biological population" (Matthews, 1960), together with that biological population, are said to constitute the ecosystem.

The Concept of the Ecosystem

The concept of the ecosystem has been a useful aid to the study of fisheries as it has facilitated an ordered approach to the problems posed; although as de Jager (1960) suggested, "many fishery workers are not convinced of the applicability of the terrestrial ecosystem concept to practical fishery problems".

Matthews (1960) stated that the marine ecosystem can be studied in three main ways:

1. By research into the physical oceanography of a given area to determine patterns of water movement and mixing.
2. By studying biotic factors under five headings so that
   (a) knowledge of the food required by a stock and the food available to it can be determined,
   (b) the predators and
   (c) the competitors of the stock can be discerned (this should also include the extent of intra-specific competition),
   (d) the diseases to which the stock might be subject can be determined, and finally,
   (e) the organic and inorganic content of the area and if possible the causes of the variations of these factors, become known.
3. Certain other factors also exert a powerful influence on the environment of any marine stock. Some of these, such as the effects of wind and sun, are not considered by Matthews to be part of the ecosystem. However, taking the definition of the ecosystem as given by Fosberg (1963) as "a functioning interacting system composed of one or more living organisms and their effective environment, both physical and biological", it is probably preferable to consider these factors as within the ecosystem. This opinion is also reinforced by Stoddart (1965), who considers the concept of the ecosystem to be more useful the more flexible it is. He states that "it may be employed at any level from the drop of pond water to the universe".

One of the problems with which fisheries biologists are always being confronted is that of the fluctuations which occur in the landings of a given species. One of their aims is to determine the causes of the fluctuations in order that, ideally, the landings of a given species for a given year can be forecast. Also it is thought that if sufficient becomes known of the underlying reasons for these changes then it might be possible eventually to even out some of the more violent alterations in the landings: alternatively much of the effort involved in hunting the fish could be saved if the fishermen definitely knew from forecasts that a certain species would not be in the area in fishable numbers until a given time.

One of the spheres of research in which the concept of ecosystem has been particularly useful is in this study of fisheries fluctuations, and as pelagic fisheries are particularly prone to fluctuations, it has been instrumental in advancing our knowledge of these fisheries.

In attempting to explain the fluctuations of the Pacific sardine, for example, the first problem which faced the biologists, and which still has not been completely solved, was whether the resource being fished was in fact one stock or more than one. More will be said of this later, but it is worth while mentioning here that it was felt that the classical morphometric and meristic approaches to the problem were not reliable for two reasons. First, the characters studied were liable to be influenced by environmental changes, and secondly earlier meristic studies had suggested greater homogeneity than had subsequently been revealed by other kinds of data such as tagging (Marr, 1957, 1960).

A Fish Population and its Reaction to Fishing

The question of stock composition of the Pacific sardine at present remains unresolved but much other work has continued simultaneously with the attempts to determine how many populations are involved, and from this a great body of factual knowledge concerning the stocks of the sardine off the coast of California has resulted. From the accumulated material several writers (e.g. Schaefer (1954) and Sette (1961)) have built up a useful picture of the reaction of a fish stock to its environment.

The picture shows a fish population under natural conditions and the reaction of this population to various fishing intensities. This is outlined below so that the reader will more readily be able to appreciate the problems involved in determining, for example, the point at which fishing intensity reaches its optimum from the point of view of the biological resource, and how this may not coincide with the economic optimum. As Morgan (1956) said, "the normal aim of commercial fishing, and any regulation of it, is to get the maximum sustained output. This is not the same thing as keeping stocks at their highest."

Starting with the assumption that the annual increase in a population is a function of population size and environmental capacity, it can be seen that a population might reach a stage where births and deaths are equal so that an equilibrium is reached and the population completely occupies or fills its environment. This is the picture without fishing. Figure 1.2 shows a possible graphical representation of the reaction of a population under varying fishing intensities. It is possible that without any fishing a population could be represented by point A in the graph. When a stock is fished, however, extra mortality is imposed thus reducing the population below the capacity of the environment. The equilibrium is upset and survival in the early stages of the life history will start to exceed the mortality giving a tendency to restore it.

When fishing is very intense and the population has been reduced to a low level the reproductive increase of the stock is near its maximum but the annual increase is low because there are few spawners. This state is represented by point *B* in Fig. 1.2. If the fishing effort is low and the population approaching the limits of the environment the annual recruitment to the stock will be quite low (see Fig. 1.2, point *C*).

The limits of the environment referred to above are largely those imposed by the relative availability of food. Inter- and intra-specific competition for the food available becomes intense so that either natural mortality is increased or reproduction is decreased or both, as the limits of the environment are approached. An environment is delimited in the physical sense, however, by the abiotic factors mentioned earlier.

Somewhere between the two extremes described above, where fishing intensity is moderate, the population level is nicely reduced so that environmental pressures are not too great,

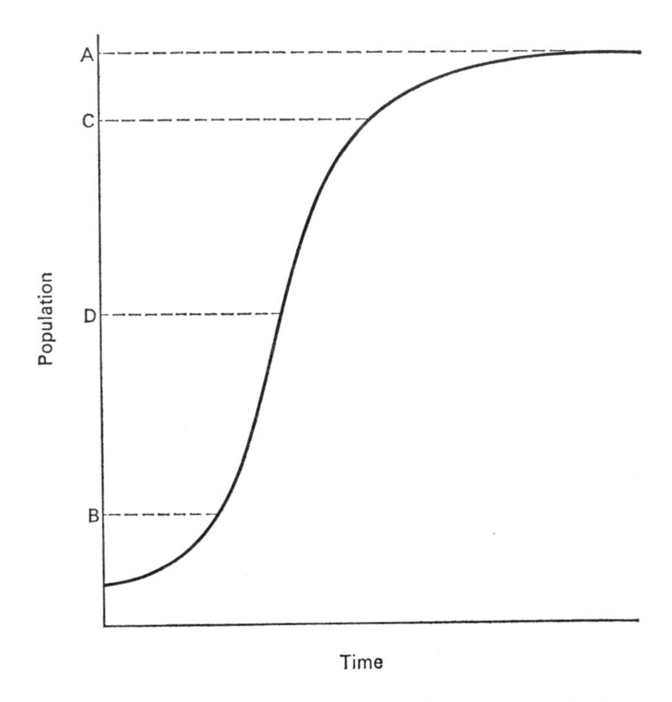

FIG. 1.2. The possible reaction of a fish population to varying fishing intensities.

the spawning stock is moderately large and the annual increase of the stock becomes maximal. This point has been represented in Fig. 1.2 as *D* and is the biological optimum, although it may not coincide with the ideal economic optimum which would probably be nearer point *C* where a greater catch could be made.

The above pictures of a stock under varying degrees of environmental pressure and fishing activity are idealized statements concerning hypothetical situations. In practice it is found that the recruitment to a fishery, that is the numbers each year reaching fishable size and age, vary widely and apparently without relation to the size of the spawning stock. This is known by a study of age-composition samples of the catch. Thus, taking a series of successive year-classes (that is the number of fish of one given age) Sette (1961) found that a tremendous annual variation existed in the numbers reaching fishable age for the fisheries he studied. Quoting his figures, four examples are given in Table 1.

TABLE 1. VARIATIONS IN RECRUITMENT AS SHOWN BY ENTERING YEAR-CLASSES FOR FOUR FISHERIES

Stock	Successive year classes	The largest was
Western Atlantic mackerel	Among 14 year-classes	15,000 times the smallest
Eastern North Pacific sardine	Among 21 year-classes	700 times the smallest
Alaska (Kodiak) herring	Among 28 year-classes	34 times the smallest
South East Alaska herring	Among 20 year-classes	13 times the smallest

Sette goes on to say that: "year-class strength must be a function of number of eggs spawned, or of survival through the egg stage, through the planktonic larval stages or through the post-planktonic juvenile stages, or a combination of these. No doubt irregularities occur in all stages, but the evidence, still regrettably scanty, points to the survival through planktonic egg and larval stages as being the most likely critical one in determining year-class strength." This is a summary of the problem of population fluctuations which will be further discussed later. It can be seen that the subjects of conservation, population studies and optimum fishing intensity are inextricably interconnected.

Fisheries Studies and the Disciplines Involved

Sometimes, when studying the history of a fishing industry, or an industry where there are no statistics covering the age and length of the catch, the research worker is confronted with figures showing the landings over a particular period. These usually show fluctuations, which may be quite marked for a pelagic fishery, and in such a case it is necessary to seek an explanation not involving population fluctuations. Quite often, as in the case of the pilchard industry of England, the major reasons for the fluctuations are found to be non-biological in origin. This is because the resource is not fully exploited and there is usually (though not invariably) a large proportion of the population which is accessible to the fishery which is not taken by the fishermen. It is then necessary to attempt to determine why the whole of the population available to the fishermen was not taken, or why the percentage which was taken in one year was larger or smaller than that taken in another year. If the technical level of the industry is not a limiting factor the causes of the fluctuations will be found to have their roots in economic or political reasons or both.

It has already been implied above that in fisheries research there is an intermingling of several disciplines, and in the studies of selected pilchard industries that follow this will become progressively more apparent. The interdependence of the various sciences can be briefly shown by an outline of the knowledge that is required in the study of a fishery and by indicating the sources of some of the aspects of that knowledge.

Thus, the biological resource is the basis of the industry. For the resource to survive to fishable size a whole host of factors considered under the title oceanography have to be satisfactory. For example, for many species the temperature of the water in which the eggs are fertilized has to be correct to within the limits of one or two degrees. If the temperature is too low development either is too slow or too prolonged to enable a good year-class to survive. Further, the processes of egg-laying and fertilization are temperature-dependent and, according to species, will not take place if the temperature be too high or too low

In the adult stage, the temperature defines the species' range, and alterations of this range can enhance or reduce the availability of a stock.

At all stages of the life history of a fish, nutriment must be available either in the form of stored reserves, or for direct assimilation. The presence of food is determined by the availability of nutrient salts, following which the dynamic series of events expressed broadly as nutrients, phytoplankton, zooplankton, can then be initiated. It is at this stage that the field of marine biology is entered.

The fish, which by the foregoing events have been brought into the fishery, now represent the apparent abundance of the species as defined by Marr (1951). Marr defined the abundance of a fish as "the absolute number of individuals in a population". The apparent abundance, he suggested, could be expressed as "the abundance as affected by availability or the absolute number of fish accessible to a fishery". He also defined availability as "the degree (expressed as a percentage) to which a population is accessible to the efforts of a fishery".

With the above definitions in mind, then, it can be stated that the fisheries scientist is interested in the available portion of the population as it is this part with which the study is now concerned. (There are some instances, of course, where it is the aim of fisheries science to equate the apparent abundance with the availability of the stock and this brings in fisheries technological research.) Technology defines the limit of the availability of a fish stock by the degree of efficiency of the boats and gear used. Then, after the fish have been caught the methods available for preservation and processing decide the extent of the marketing channels to which the product can be subjected.

The marketing of a product is itself concerned with a multitude of different studies such as transport, advertising and other more specialized methods of consumer education required when new products or new species are being sold. All these are, in turn, basically affected by the laws of supply and demand. The economics of operation play a major role in determining the extent, in terms of duration of fishing trips and selecting grounds to be fished, of fishing operations, the methods of fishing and processing used and the scope of marketing operations.

It is hoped that the preceding pages have indicated the range of subjects which form an integral part of the study of a fishery. In the next chapter the scope of the book in relation to the global distribution of pilchard species will be defined and the treatment to be afforded to the species chosen for study will be outlined.

CHAPTER 2

GEOGRAPHICAL DISTRIBUTION OF SARDINES ZOOLOGICAL CLASSIFICATION, ETC.

FISH known colloquially as pilchards or sardines are classified biologically into three distinct genera and at least eighteen different species. They are found around the shores of every continent and, although they are fundamentally warm-water species, their global distribution extends between the latitudes of 60°N and 50°S. Figure 2.1 shows the distribution of sardines and indicates that there is a profusion of species in the waters off India, eastwards to the coast of China, and in the offshore regions of the archipelagos of Japan and Indonesia. In other areas single sardine species may predominate, perhaps to the exclusion of the others. For example, in the English Channel *Sardina pilchardus* is the only sardine species found; and the stocks of *Sardinops caerulea* off the coast of California have been the sole supply of the Californian sardine industry.

Little is known about many of the species of *Sardinella* which form the basis for the fisheries of the Persian and Oman Gulfs and of the Indo-Pacific regions. However, important work has been done concerning these species, and papers by Kwan-Ming (1960), Tokyo (1960), Soerjodinoto (1960), Nair (1960) and Ronquillo (1960) contain valuable information on their biology. Apart from the landings of India and Japan, quantities caught by other countries of these regions have been small or not included in the statistics (F.A.O., 1964e). Further information about the landings of these countries is given in the Appendix.

In this book, the European pilchard, the Pacific sardine and the South African pilchard have been chosen for detailed study. These three fish species are biologically similar and have been the basis for fisheries which have developed in remarkably different ways. The European pilchard has been fished from the English Channel off the coasts of Cornwall for hundreds of years and has recently been the subject of much local controversy. This Channel resource is generally considered to be underexploited (Cushing, 1957), and since the end of the Second World War has been the object of several scientific investigations which have attempted to show how the resource could be more fully utilized (Hodgson and Richardson, 1949; Bridger, 1965). The biology, technology and economics of this industry will be dealt with in detail, and the account of this industry is presented as a case study.

It was in the second decade of this century that the Pacific sardine was first fished on a commercial basis. The development of the industry was extremely rapid and reached its climax in the 1930's. Since then it has declined and in 1963 reached a level almost as low as that of the English industry. This resource was chosen for study, therefore, because it shows the rise and fall of an industry—a feature which is characteristic of pelagic-based fisheries in both short-term and long-term cycles of fluctuations. It is an exceptionally well-documented industry and much research has been done to decide in retrospect whether the overall decline since the 1930's could have been averted. The question still seems moot and will be discussed in more detail later.

The South African pilchard is of interest because it is the fecund resource on which the pilchard fisheries of South and South West Africa rely. These two industries have developed very recently, since the Second World War. In fact, it was not until 1951 that commercial fishing for pilchards began in South West Africa. These are industries which are significantly expanding their catches, and which are now fishing the biological resource at what may be near the maximum possible before overfishing causes a decline. Much research is being done in an attempt to prevent an avoidable abuse of the resource, and this will be discussed later, but the rapid rise in catches and fishing effort in the area inevitably calls to mind the history of the California industry.

Thus, these three industries are examples of fisheries based on:

1. an underexploited resource;
2. an overfished resource, or at least a much reduced resource;
3. a flourishing resource being heavily exploited, respectively.

Biological Nomenclature

Duhamel du Monceau (1772) considered that the European pilchard should be placed in the genus *Clupea*, as did Day (1880). However, Walbaum (1792) is credited with first calling the fish by the specific name of *pilchardus* and he suggested *Sardina* as a generic name. Several other generic names have been proposed but recently *Sardina pilchardus* has been accepted as the official name by the International Commission on Zoological Nomenclature as a result of a paper by Wheeler (1964). It should be borne in mind, however, that *Clupea pilchardus* was used for many years and that as a result many writers still use this name. In fact, the majority of the papers concerning the work on the pilchard in the English Channel contain the old name, possibly because much of the work was done before the publication of Wheeler's paper. The previous writings of the present author always used the name *Clupea pilchardus* because of its frequent use by other workers on the species in spite of the fact that for many years the Plymouth Marine Fauna listed *Sardina* as the generic name (Marine Biological Association of the United Kingdom, 1957); and also because *Clupea pilchardus* was thought to be less confusing as it incorporated only one latinized common name and this is the name by which the fish is known in England; and as Russell pointed out "some common names for fish are almost as valid as their scientific names, and they are certainly more stable". (Marine Biological Association of the United Kingdom, 1957.) However, with the official acceptance of *Sardina pilchardus* this will now be used for the European pilchard throughout this book. It is important to note that the sardine of commerce is the young, usually sexually immature, stage of the pilchard. That is, up to the age of about 3 years *Sardina pilchardus* is the "sardine" after which it becomes the "pilchard" of commerce.

For the Pacific sardine, there has in the past been some discussion concerning the correct generic name. However, in 1929 Hubbs suggested that *Sardinops* should be the generic name, and although *Sardina* and *Sardinia* persisted for some years after this, *Sardinops* became gradually to be accepted. The specific name is *caerulea*. The nomenclature at present accepted, *Sardinops caerulea*, is attributed to Girard (1854), although the generic name that he suggested was *Meletta*. The earliest recorded suggested nomenclature for the Pacific sardine seems to have been *Clupea sagax* by Jenyns (1842).

It has been suggested (Svetovidov, 1952) that five species of *Sardinops* are so similar as to merit placing under one name. There is evidence of a "lumper" at work here! (See Kerkut, 1960.) These are:

(i) the Pacific sardine, *Sardinops caerulea* (Girard), 1854;
(ii) the Japanese sardine, *Sardinops melanosticta* (Temminck and Schlegel), 1846;
(iii) the Australian and New Zealand pilchard, *Sardinops neopilchardus* (Steindachner), 1879;
(iv) the South African pilchard, *Sardinops ocellata* (Pappé), 1853;
(v) the Chilean sardine, *Sardinops sagax* (Jenyns), 1842.

Svetovidov considered that the differences between the species might merit sub-specific rank and he proposed that these should be sub-species of *Sardinops sagax* (Jenyns). Work done at the Scripps Institute of Oceanography (Ahlstrom, 1960) supports Svetovidov's contentions, but takes his theory further, the proposal now being that all the above species should be known as *Sardinops sagax* (Jenyns) with (iii), the Australian and New Zealand pilchard, having a sub-specific rank.

The position regarding the official nomenclature of the sardines and pilchards of the world is thus seen to be a rather vexed question. It will, therefore, be understood if it is stated that at present the South African pilchard is classified as *Sardinops ocellata* (Pappé) and that in the past there has been a certain amount of synonymity with the generic names of *Sardina*, *Arengus*, and with the binomial *Clupea sagax*.

The names that will be used throughout this book then, for the three main species under consideration, are:

(i) *Sardina pilchardus* (Walbaum), 1793, for the European pilchard;
(ii) *Sardinops caerulea* (Girard), 1854, for the Pacific sardine; and
(iii) *Sardinops ocellata* (Pappé), 1853, for the South African pilchard.

Supra-generic Classification

Several alternative systems are available for the overall classification into ranks above genus. A broad outline of the classification of clupeoid fish is given in Young (1962). More detailed accounts relating to teleostean fish as a whole (Greenwood *et al.*, 1966), or relating particularly to the Isospondyli (Gosline, 1960), are available. Greenwood *et al.* (1966) recommend a major revision of the teleostean fish along phylogenetic lines and consequently some genera, formerly included in the order Clupeiformes have been removed from this order and placed in others. A workable classification based on the above references could be represented as follows:

Phylum	Chordata
Sub-phylum	Vertebrata
Super-class	Gnathostomata
Class	Osteichthyes (Actinopterygii)†
Sub-class	Neopterygii
Division I	Teleostei
Super-order	Clupeomorpha
Order	Clupeiformes (Isospondyli)†
Sub-order	Clupeoidei
Family	Clupeidae (Clupidae)†

† Names in brackets are common synonyms.

The characters by which a particular fish is assigned to the Clupeomorpha category of Greenwood *et al.* are listed below.

"1. Silvery, compressed fishes, usually marine, with caducous scales.
2. Branchiostegals numbering as high as 15, but usually fewer.
3. Intercranial diverticula of swim bladder forming bullae within the ear capsule.
4. Mesocoracoid arch invariably present.
5. Hypurals on one to three centra.
6. Cephalic lateral-line canals extending over operculum; usually no lateral-line pores on trunk.
7. Recessus lateralis present." (Greenwood *et al.* (1966), p. 350.)

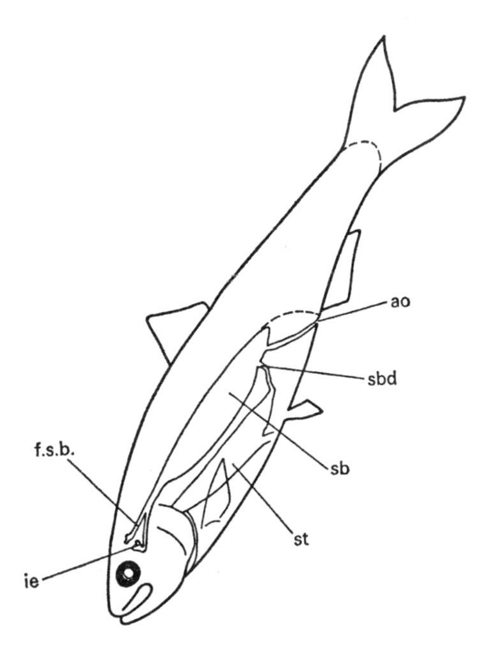

FIG. 2.2. The linkage of ears and swim bladder in herring-like fish. ao, duct from swim bladder with opening near anus; fsb, Y-shaped forward extension of swim bladder; sb, swim bladder; sbd, duct from sb to stomach (st); ie, ears.

Characteristics of the Family

On the whole the members of this family are of generalized structure, but the following points are usually considered characteristic, although all these features are not necessarily found in all species.

1. The supra-occipital separates the small parietals.
2. There is generally a superior temporal fossa (between the frontal and parietal), and a pre-epiotic fossa. Also pro-otic and pterotic bullae are usually present. These enclose diverticula of the gas bladder and are referred to in 3 above. (See also O'Connell, 1955, and Fig. 2.2.)
3. An auditory fenestra is usually present.
4. The eye-muscle canal is generally open behind.

5. The upper jaw is of variable structure but the margin is formed by the premaxillaries or the maxillaries or both.
6. The premaxilla and maxilla may be firmly joined or the latter may articulate with the posterior ethmoid.
7. Teeth may be present on the premaxilla or maxilla or both.
8. One or two supra-maxillae are usually present.
9. The opercular bones are well developed.
10. There is a remarkable development of intermuscular bones; epineurals and epi- and ad-pleurals. Usually the pleural ribs are joined below by a series of median V-shaped scales to form complete hoops.

About 300 clupeid species are known to live at the present day (Marshall, 1965). These are mostly marine but none is a deep-sea form. They have a very wide geographical range, as illustrated by the sardines being considered here, and some are anadromous, ascending rivers to spawn, for example the allis shad (*Alosa alosa* L.) and the twaite shad (*Alosa fallaex* Lac.).

It can be seen that biologically the fish to be considered here belong to a very interesting group which have apparently been able to adapt themselves well to their environment becoming dominant species in their class. It will be shown, however, that in one case at least Man has upset the balance of the environment to an extent which may be the prime cause of their being removed from this dominant position. In the areas where these species are exploited they have a significant, economic importance and will continue to be important in the future. This is because with the increasing world demand for fish, fish meal and, in the future, for fish protein concentrates, they are a valuable source of protein and, provided that the resources are not overfished, should continue to be so.

In the next chapter some of the oceanographic and biological factors impinging on fishery resources will be considered in more detail.

CHAPTER 3

FACTORS OF PRODUCTION IN THE SEA

As an environment the sea is in many ways ideal. Those plant organisms growing in it are bathed in a solution containing the nutrient salts and gases required for their growth, and which removes from them the waste products of their metabolism. For animals, the sea transports the gases of respiratory exchange to and from their respiratory surfaces, and supplies organisms for their nutriment. However, the environment is liable to appreciable changes. To find out how the environment alters it can be analysed, and several factors may be selected for further, more detailed, study. Although these have to be dealt with individually, it is important to remember that they are, on the whole, closely interrelated so that a change in one can bring about a change in another which may cause a rapid alteration in the environment. Some of these are dealt with below.

Abiotic Factors

1. *The Distribution of Salinity*

Water from various areas can be identified by its salinity concentration. Furthermore, when water moves from one area to another, it tends to keep its typical salinity and can thus be identified in its new position as having originated elsewhere. For example, in Fig. 5.8 there is an illustration of the typical salinity distribution in the English Channel and North Sea during winter; an area of water in the mouth of the Channel is shown as being more saline (35·3‰) than the surrounding water. It is known that this is a projection of water from the Atlantic. Although salinity concentrations do not usually undergo marked changes in any given area and are, therefore, within the salinity-tolerance range of most organisms, their analysis has greatly aided the study of water movements.

2. *Water Movements*

Fundamental water movements are effected by several means, of which (i) wind action, (ii) differences in temperature between two different bodies of water, (iii) differences in salinity between different bodies of water, are the most important. These movements are, furthermore, substantially modified by the rotation of the earth and the configuration of the land.

There is a large body of literature in existence which deals with ocean currents of the world, but discussion of these is beyond the scope of this book. (More detail can be found, for example in Colman (1950, pp. 40–84) or Sverdrup *et al.* (1942, pp. 431–761).) It should be noted here, however, that one of the most important results of the interaction of different water masses is the occurrence of up-welling currents in certain places. Steady off-shore winds, such as the Trade Winds, also create areas of high primary production through up-welling necessary to replace water driven off-shore. This is because these up-wellings bring with them large concentrations of the nutrient salts, nitrates and phosphates, which play a very important part in the biological dynamics of the sea.

3. *Influences of Temperature*

Direct solar radiation is the main heat source of the sea. The amount of radiation received from the sky is also important and varies with the position of the sun and the degree of cloudiness. For example, with a clear sky and high sun, about 85 per cent of the radiation is direct and about 15 per cent is from the sky. However, if the sun is low, say 10 degrees above the horizon, the amount of radiation from the sky may be as much as 40 per cent. There are several other minor heat sources (see Sverdrup *et al.*, 1942).

The temperature cycle of the sea taken over a year in any one place at a given depth varies each year and is dependent, not only on solar radiation, but also on the meteorological conditions, the amount of mixing which occurs and on any inflow of water from other parts. In the northern hemisphere the temperature of the seas normally starts to increase in April or May and a warm upper layer, which may reach 20 m in depth, is formed. Below this layer, sometimes referred to as an epithalassa, the water is cooler and more or less isothermal. The boundary between the epithalassa and the cooler water below is called the thermocline. Once formed, the thermocline is remarkably stable, as the water below it being cooler is, therefore, denser, and intermixing is negligible.

At the beginning of September (in the northern hemisphere) there is not usually any further gain in heat. The occurrence of a few consecutive days of high winds begins to break up the thermocline and to increase evaporation, so that soon the temperature of the sea has fallen considerably and becomes more or less uniform from the surface to the sea bed. Other causes of heat loss are radiation, absorption and conduction outwards into the atmosphere. Apart from gales, the thermocline begins to break down due to penetration of convection currents into the hypothalassa. These occur because of the cooling of the epithalassa. In the English Channel the major loss of heat has occurred by the beginning of December, although the lowest temperatures are usually not reached until March. Apart from the effect of temperature changes on the sea already mentioned, namely that part they play in the building up and breaking down of the thermocline, the temperature itself has marked effects in several ways on the animal life of the sea. Two of the most important of these are:

(i) the way in which the breeding is affected, and
(ii) the way in which distribution of a species may be governed by temperature.

(i) *Breeding*. For some marine animals it has been found that the temperature of the sea determines the start of the breeding season, and may also determine when it ends. In considering the effect of temperature on the breeding of marine animals it has been shown that there are two basically different types:

(a) those which can breed provided that the temperature remains above a certain figure—a group of animals to which pilchards belong;
(b) those which can breed provided that the temperature is less than a certain figure.

A third group of animals exist which has such a wide temperature-tolerance range that the animals in it continue to breed for most of the year. It is generally found that such animals have a very wide distribution. Thus, *Ciona intestinalis* (a sea squirt) breeds all the year round at Naples and nearly the whole year round at Plymouth (Orton, 1920).

(ii) *Distribution.* It is a natural result of the effect of temperature on reproduction that temperature will also affect the distribution of many species. From this fact three conditions follow. First, ubiquitous forms must have a wide temperature tolerance for breeding. Secondly, high latitude forms in relatively low latitudes should find their breeding period when the temperatures are relatively low for the area. The third is the converse of the second, namely, low latitude forms in relatively high latitudes should only breed in the summer of that area. Warming up of a given area would probably favour a genus of group (a) in competition with a genus of group (b). Cooling would have the opposite effect. For example, it seems (Southward, 1963) that the increasing Channel temperature has possibly favoured the pilchard, *Sardina pilchardus*, especially when considered in the light of competition with the herring, *Clupea harengus*; the former belonging to group (a) and the latter to group (b).

4. *Changes in Concentration of Salts of Major Importance*

The two most important elements for the maintenance of life in the sea are nitrogen and phosphorus. To be of use to plants the nitrogen must be in the form of nitrates, nitrites, some amino acids or ammonia, and the phosphorus in the form of phosphates. Some organic phosphorus-containing salts are also of use. It has been found that the nitrate: phosphate ratio in the sea is about 9:1 by weight and variable over a limited range only (Cooper, 1937).

Estimations of phosphorus in water samples in an area such as the western English Channel vary considerably, from as low as 10 mg phosphorus per cubic metre, to double or more than double this concentration. This is low when compared with many other areas. For example, the water in the Gulf of Maine (off the north-east coast of the United States of America) has, on an average, three times more phosphorus in the water than that off Plymouth (Riley and Bumpus, 1948).

It has been suggested (e.g. by Harvey, 1950) that there is a direct relationship between the maximum winter phosphate concentration of an area and the potential fertility of that area for the following year. However, this relationship has not always been found to hold good. For example, Southward (1962) has indicated that although there has been a drop in the winter phosphate level in the western English Channel, it is doubtful if there has been a corresponding decrease in the fertility of the area, if indeed a decrease can be shown at all. Cushing (1953) has suggested that the relationship is more complex than this, being linked with the grazing of the phytoplankton population and the rapid recycling of nutrients which this allows. However, a change in the concentration of phosphate in a given area denotes a change in the supply of nutriment to the phytoplankton which may bring about a change in the size of the phytoplankton population. This will in turn affect the size of the zooplankton population as the phytoplankton and zooplankton populations are interrelated.

There are certain other ions which are essential for the growth of plants (apart from the other major elements such as carbon, oxygen, hydrogen, sulphur and potassium) which need only be present in trace amounts. These include the ions of magnesium, iron, manganese, copper, etc., which are often found to be the central element of a complex ion or molecule as, for example, the magnesium ion in the molecule of chlorophyll. These ions are often referred to as growth promoters. There are also some plants with special growth requirements such as the need of diatoms for silica.

Biotic Factors

In this section phytoplankton and zooplankton production is considered in relation to the abiotic factors and the relationships between these two main groups are discussed.

Phytoplankton

The conditions necessary for the production of plant tissue in the sea are summarized below. Plant production is dependent on:

1. The rate of regeneration of nutrients from dissolved organic compounds, particularly within the photosynthetic zone.
2. The quantity of nutrients rising into the photosynthetic zone from below. These enable growth to continue. The extent to which the stock can grow is, however, dependent on:
3. The population density of zooplankton grazing the stock.
4. The amount of reduction in the number of growing cells caused by their sinking below the photosynthetic zone.
5. The depth of the photosynthetic zone.

On this factor also, and on 6, 7 and 8, depends the growth rate of the growing stock of plants.

6. The effect of temperature on the respiratory rate of plants.
7. The concentration of nutrients within the photosynthetic zone, which correlates positively with 1 and 2.
8. The concentration of certain growth-promoting ions, such as those of manganese, magnesium and iron, present in sea water (Harvey, 1950).

A method frequently used to estimate the amount of plant material present in a given area is to determine the amount of chlorophyll present in the water. (This probably usually gives an optimistic estimate because of chlorophyll present in the herbivores.) The extracted pigments are compared with an arbitrary standard of colour known as one unit of plant pigment (abbreviated as U.P.P.). Harvey (1950), using his own experimental results, as well as those of others, estimated that for a mixed plant community growing in the comparatively nutrient-poor waters of the English Channel, one U.P.P. contained, as a rough approximation, about 0·016 mg of dry, ash-free organic matter. He was later able to estimate that, for the Plymouth area, total production below a square metre for one year (1949 in this case) would probably be not less than 120–200 g, or 0·4–0·55 g/day. This compares unfavourably, as would be expected, with the results, given by Riley (1941) and converted into comparable units by Harvey, for the Gulf of Maine, which produced 270 g of organic matter below a square metre per year, or 0·8 g per day. It was found that the production in the first half of the year was greater than for the second half.

Zooplankton

Zooplankton is largely composed of plant-eating juvenile copepods and their nauplii, but it has been stated by Bainbridge (1953) not to be a distinct group, but rather "a loose assemblage of animals with clearly graded powers of locomotion". Copepods have a daily increase of from 7–10 per cent of their weight. Experiments with *Calanus finmarchicus* and a mixed community of crustacean plankton indicated that 4 per cent of the carbon in their tissue was used for their daily respiration at the summer temperature of the sea. Therefore, a population below a square metre of the sea containing an average of 1·5 g of organic matter will require to assimilate 4 per cent of vegetable organic matter to offset respiratory losses each day; that is 0·06 g is required for this purpose. A further 7–10 per cent (or up to 0·15 g) is needed to build new tissue. To give approximate values, it can, therefore, be said that from plant to small zooplankton about 30 per cent of the assimilated energy is lost in respiration and about 70 per cent converted to animal tissue.

The size of the zooplankton population in any one area at a given time tends to be directly dependent on the phytoplankton. It has been demonstrated, on several occasions, that although both phytoplankton and zooplankton are very uneven in distribution there tends to be an inverse relation between the two in any one limited area. Hardy and Gunther (1935) were the first to emphasize this, and theories to account for it have been proposed by Harvey (1934), Steemann Nielsen (1937), Lucas (1936 and 1947), as well as Hardy (1936). The hypotheses presented by these authors have been reviewed and tested by Bainbridge (1953) who concluded that zooplankton tend to congregate in localized areas of high concentrations of phytoplankton. Because the phytoplankton is tasteful to the zooplankton the former soon diminish in number. There then exists a large concentration of zooplankton and only a small concentration of phytoplankton, whereas previously the reverse situation has been observed.

Annual fluctuation in the density of zooplankton populations occur, and these can be caused by several factors. For example,

1. different bodies of water passing through the area containing different amounts of nutrients could alter the size of the phytoplankton population—i.e. the source of nutriment of the zooplankton population;
2. differing rates of regeneration of nutrients from year to year would have the same effect;
3. predatory action by migratory pelagic fish and others varies from year to year;
4. the proportion of carnivorous plankton itself possibly also changes from year to year;
5. turbulence has been found to affect the size of zooplankton populations;
6. changes in other factors such as temperature and light affect the zooplankton either directly or via the phytoplankton population.

Productivity of the Sea

The ultimate measure of productivity in the ocean is the amount of carbon fixed by photosynthesis over a given area for a stated time. Taylor (1955) quotes an estimate (carried out by Riley) of the average carbon fixation for a square kilometre of sea per year which he compared with the value for the land. For the sea he stated that about 340 tons of carbon per square kilometre per year is fixed. On land the figure was 160 tons/km^2 per year. Since only 29 per cent of the earth's surface is land, the total photosynthesis of the land area of the earth must be 20 billion tons compared with 126 billion tons over the parts of the world covered by the sea.

When a zooplankton organism feeds on a member of the phytoplankton and the zooplankton is eaten by another larger animal, it is reckoned that in these stages there is an average of only 5–10 per cent efficiency. Or, for every 5–10 lb of animal growth, 100 lb of food is needed. A simplified food chain might be represented as follows:

10,000 lb diatom ≡ 1000 lb copepods ≡ 100 lb herrings ≡ 10 lb mackerels ≡ 1 lb tuna.

Using such reasoning the quantity of carbon per square mile incorporated in useful fish has been estimated. However, when this was compared with the actual commercial yield of a given area it was found that there was more carbon incorporated in the commercial fish alone than was theoretically possible for the area under consideration.

This leads to the conclusion that the natural production of the sea must be more efficient than has hitherto been thought. Therefore, either the synthesis of vegetable matter must be more efficient, the system of food transformation from one stage in the food chain to another must be more efficient, or the number of stages might be less than had been assumed in the calculations. This is the same problem, expressed in applied terms, as that mentioned earlier dealing with the winter phosphate maximum and potential production of an area. It can be seen, therefore, that a distinction has to be drawn between the potential production of an area and its actual production. Earlier estimates of potential production were based on static considerations, i.e. an attempt to relate a continuous process with a single estimate of one factor of production. Actual production is a measure of the standing crop of phytoplankton and its production rate over a period of a year. The two differ in three important respects. First, there is a loss of production between trophic levels in the food chain, which is either incorporated into new material or lost as energy. Secondly, due to death and decay, it is possible for the phosphates to go through the production cycle more than once per year. Thirdly, life cycles of certain animals, fish included, are not completed in one year and they thus represent a reservoir of energy. In all three cases, actual production may be greater than an estimate of potential production, should these factors not be taken into account.

Data concerning food transformation efficiency are sparse. Most are concerned with the energy relations of the production cycle and from these it was reckoned that each transformation was about 10 per cent efficient, as mentioned earlier. More detailed knowledge confirms that this figure was a useful average, because it has been estimated that in the larval juvenile stages of fish and invertebrates the efficiency of food transformation is as high as 40 per cent, declining to 15 or 10 per cent in adults and being less than this in old age. Thus, with respect to the energy cycles, this determination of efficiency has, in effect, been relying on a negative principle, that is a loss of energy.

Quantitating the connections between the biotic factors in a given food chain is experimentally very difficult and some stages have not yet been adequately studied. Those areas that have been studied have not been applied fully to resource management. The four points outlined below present one possible way of approach to the problem.

(i) A ratio between the nutrient salts present in an area and the phytoplankton produced in that area should be determined.
(ii) The relationship between phytoplankton and zooplankton production of the area must be ascertained.
(iii) It is also necessary to determine the relationship between the zooplankton and any intermediary in the production chain leading to the fish being considered. This stage may not exist or there may be several stages of this type depending on the stock being investigated.

(iv) A quantitative relationship (probably based on energy transformations) should be assessed between the zooplankton production (dependent on the phytoplankton) and the intermediary production (dependent on the zooplankton) and the fish (end product).

Fleming (1939) devised an equation to estimate the effect which the zooplankton population had on the diatom population by grazing. The equation was

$$\frac{\Delta P}{\Delta t} = P[a - (b + ct)],$$

where a change in the diatom population with respect to time is dependent on a, the rate of multiplication of the population; b the initial grazing rate; and c the change of grazing rate due to changes in the zooplankton population. The equation assumes that:

(i) the rate of multiplication of phytoplankton is constant,
(ii) an animal grazes a fixed proportion of the diatoms per day (that is, that a unit volume of water is filtered per day),
(iii) there is no loss other than by grazing and
(iv) the effects of mixing due to turbulence are negligible.

Cushing (1953) developed this approach further when he considered a *Calanus* population in terms of its reproduction (which is dependent on the production of plants) and also in terms of its death rate (which is partly dependent on the grazing of the herring). He found that there is a two-way dependence between all of the trophic levels except the last, where the predatory effect of man on the herring population was taken into consideration. Cushing demonstrated also that when considering the effect of fishing activity the equations for fish and drifter differ from those for plants and herbivorous zooplankton in that the numbers of fish and drifters in any one area are dependent also on the numbers of these factors elsewhere.

When more is known of the details of the trophic stages outlined earlier it will be possible to say concerning a fishery, for example,

(a) what increase in the weight of the fish stock was due to the grazing on zooplankton or some intermediary;
(b) whether there is enough plankton for the young fish and the intermediaries, or whether the intermediaries or other competitors deprive a certain percentage of the stock of young fish or food thus reducing its size;
(c) by how much a fish stock is reduced by predators.

This will lead to much more efficient husbandry of marine fish stocks as such questions may lead to others. For example, considering (c) above makes one realize that to answer this point it is necessary to know all the predators of a stock. It is likely that some of these will also be of possible commercial value themselves, possibly of more value than the first stock being investigated. The predators will, of course, be at a higher trophic level than the first stock. On closer consideration it may appear that under certain economic conditions it would be better management of resources to leave the fish which is at the lower trophic level to be eaten by the predators. The predators can then be fished when they have reached their most useful size for exploitation. This could possibly lead to a fish at an even higher level being somewhat deprived of food. Thus, the problem emerges as a complex one but of a type which may well have to be considered in the future.

Similarly, if it were known that two stocks were feeding in the same area and one required more zooplankton per pound of increase in weight, then it would be economic sense to fish that stock in preference, thereby leaving more zooplankton for the other stock. This other stock would then flourish and could in turn be exploited. Finally, it is useful to know the growth curve for commercial species. It is known, for instance, that in some stocks it is economically advantageous to overfish deliberately the older age groups, leaving a population which can accomplish the maximum growth for the food eaten, without having competition for food from non-productive (in terms of edible flesh) old fish. Conversely, it might be better to remove a large proportion of the stock of young fish. This would increase the ratio of "available food: numbers of fish", thus enabling the remaining population to grow faster.

More research into these problems will eventually provide data enabling the dynamic relations of the food chains of commercial fish to become better known. The application of this knowledge will make for a more efficient and economic use of the resources of the sea.

PART II

THE PILCHARD INDUSTRY OF ENGLAND

CHAPTER 4

THE BASIC RESOURCE OF THE ENGLISH PILCHARD INDUSTRY

THE pilchard has been fished off the coasts of Devon and Cornwall for at least 400 years and the stock has generally supplied a catch which has averaged at between 3000 and 5000 tons annually during this period.

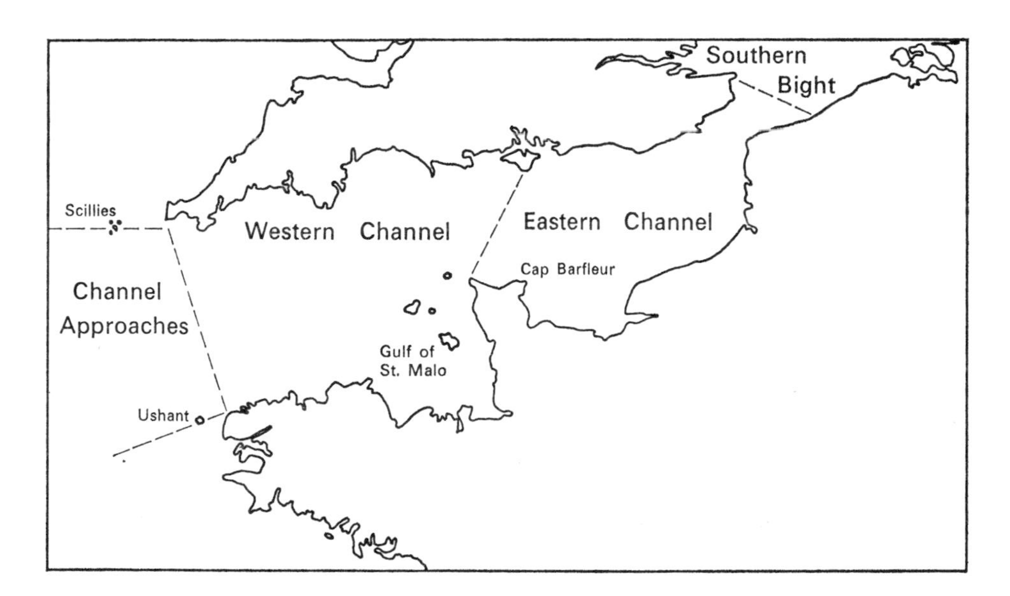

FIG. 4.1. The English Channel, showing the area of the Western Channel and its approaches.

The main area of study connected with the Cornish pilchard is shown in Fig. 4.1 as the Western Channel. However, the whole of the English Channel will be considered at those times when either water movements, from the eastern region of the Channel to the west, or movements of fish, from the western region of the Channel to the east, make it necessary to do so.

The pilchard has been confused with the herring, and sold as such from time to time, and its younger stages have been muddled with the younger forms of two of its close relatives the sprat and the herring. A suggested classification of the pilchard has already been given, and various methods by which it may be distinguished from its clupeoid relatives in British waters, both in the larvae and adult stages, follow.

Using the method devised by Lebour (1921a) the larvae are distinguished by the numbers of myotomes present. The total number differs as does the number between the operculum and the anus. This classification for an 8–10 mm larvae is given in Table 2.

TABLE 2. SHOWING THE NUMBER OF MYOTOMES OF THE BODY AND BETWEEN THE OPERCULUM and ANUS IN SPRAT, PILCHARD AND HERRING LARVAE

Total number of myotomes in body	Number between operculum and anus (trunk)	Species
46–48	Usually 37	Sprat
51–52	Usually 42	Pilchard
56–58	Usually 47	Herring

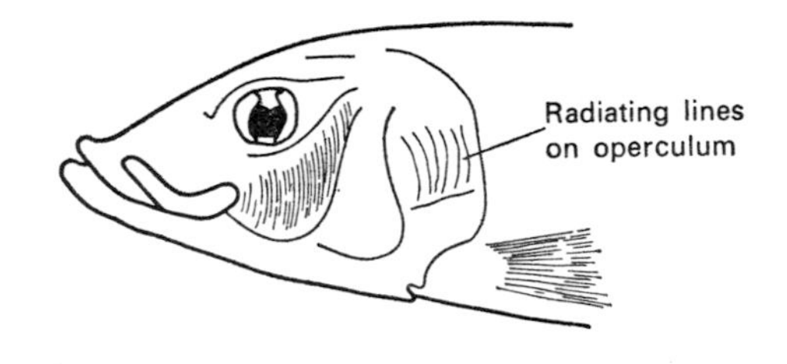

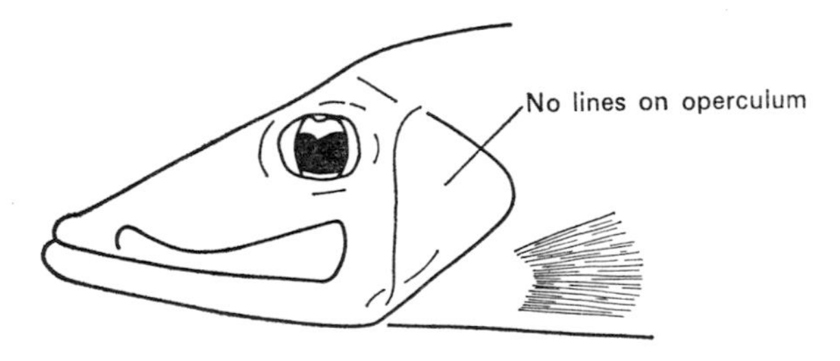

FIG. 4.2. The heads of a pilchard and herring compared.

For larger forms other factors are used. Some of these are the time that the pelvic fins appear and their position in relation to the pylorus; the length of the tail is also sometimes used as an identifying feature. It was found that for any particular length the stage of development of the herring was more retarded than for the other two species.

The adults of the herring and pilchard (*C. harengus* and *S. pilchardus*) are easily distinguished from the adult of the sprat (*C. sprattus*) on size alone. The adult herring ranges from 20 to 28 cm in length, the pilchard from 18 to 23 cm and the sprat reaches 6–8 cm in length. The adults of the two larger species can be distinguished as the herring is much bluer in appearance than the pilchard, which is greenish. Also the herring is perfectly smooth along the belly but the pilchard has scales along the undersurface which make it rough to the touch. The scales of the pilchard are unmistakably larger than those of the herring, and the pilchard has distinguishing marks on the operculum (see Fig. 4.2). In this figure lateral views of the heads of herring and pilchard are compared and the marks on the operculum of the pilchard can be seen.

The position of the dorsal fin in the three species varies in relation to the head and the tail. If the herring is suspended by the dorsal fin it is evenly balanced, but if the pilchard is held by the dorsal fin then the fish hangs with the tail lower than the head. Conversely, the sprat held in a similar way balances with the head lower than the tail.

When such characters are borne in mind all these species can be easily distinguished and there is no reason why the confusion, which has sometimes been manifest in the marketing of these fish, should persist; although the younger stages of pilchard and herring will probably always be difficult to distinguish from the sprat when mixed with them.

Distribution

The race of pilchard found in the English Channel extends west as far as the south of Ireland, and south as far as Gibraltar. The fish of this group are distinguished from the sardines of the Mediterranean and the west coast and north-west Africa by a higher vertebral count, although the averages of the different species or sub-species differ only slightly. The northern species has a vertebral count of approximately fifty-one (Hodgson, 1957). They are still called by the same name and literature dealing with the southern species seems to have used *Sardina* rather than *Clupea* in the past.

In the Mediterranean and round the north-west coast of Africa there is an intermingling with another closely related genus, *Sardinella aurita*. This genus has a considerably lower vertebral count (about forty-eight) and extends down the whole of the west coast of Africa. *Sardina pilchardus* does not extend its range further south than about 20°N, and the northern and southern forms mentioned above may intermingle in the Western Channel. Apart from a few records of pilchards from the Minch, the northern limit of *S. pilchardus* seems to be the southern North Sea.

Food and Feeding Habits of the Pilchard

The plankton consists of all those animals or plants which are at the mercy of currents and wind and which, therefore, drift about the oceans. From the point of view of food for the pilchard, plankton falls into three categories.

1. That which is never used. This category includes animals which have never been found in the stomach of the fish. They are animals which are avoided for some reason or other, for example, they may be able to sting, like jellyfish and sea gooseberries.
2. A secondary food of the pilchard which includes many phytoplankton species and some Metazoa associated with the phytoplankton.
3. The preferred food of the pilchard which is the highly nutritive zooplankton. The most important examples, to judge from stomach contents, are the copepods *Calanus finmarchicus*, *Pseudocalanus elongatus* and *Centropages typicus*, and larvae of decapod Crustacea (Hickling, 1945; Swithinbank and Bullen, 1914).

In the category 3 above, the word "preferred" is used as it appears that the pilchard is able to select its food. This conclusion was reached after stomach contents had been compared with plankton samples taken in the vicinity of actively feeding fish and it was noticed

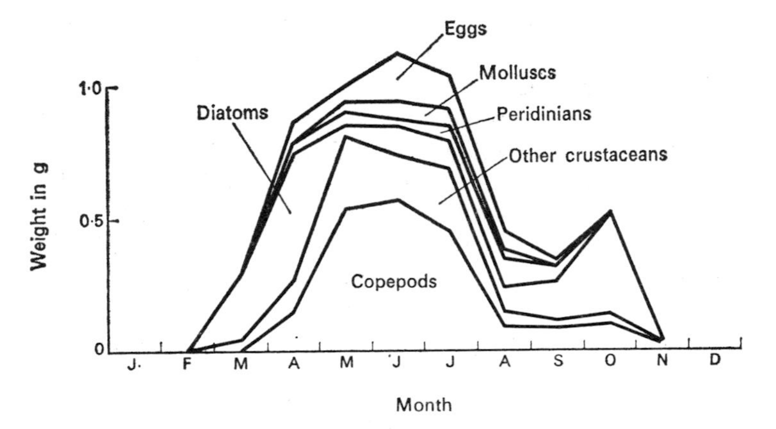

FIG. 4.3. Principal constituents of the English pilchard 1935 through 1938.

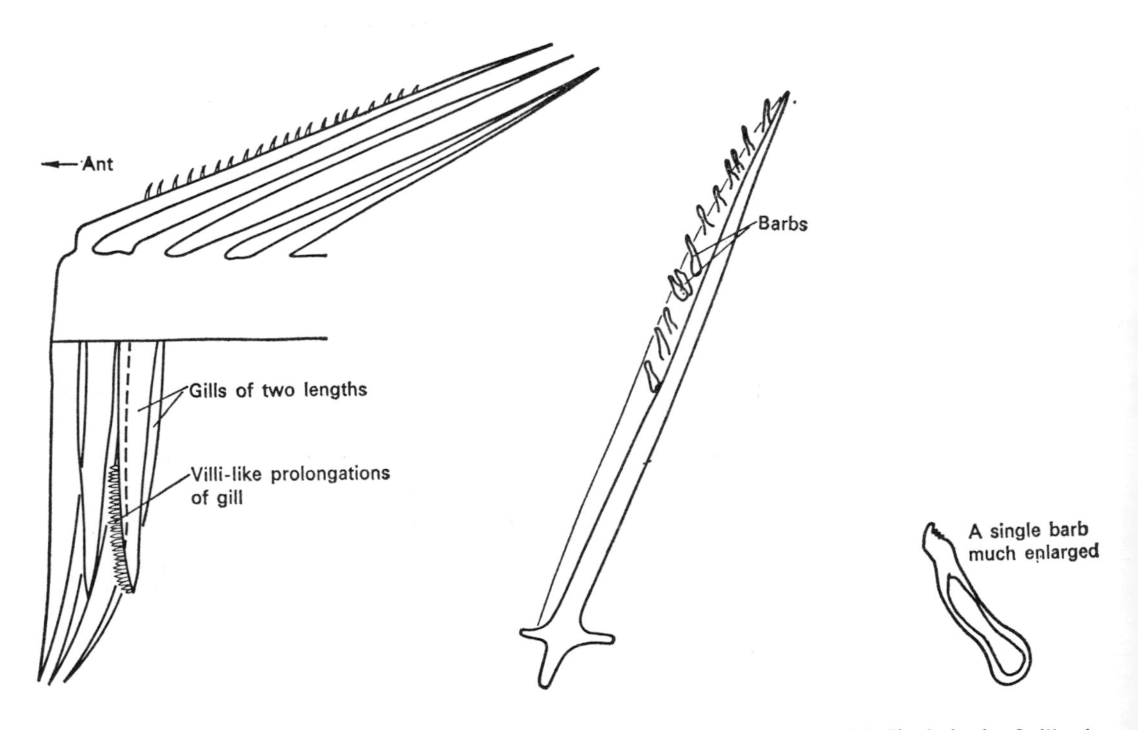

FIG. 4.4. Gills and gill rakers of *Sardina pilchardus* (×4).

FIG. 4.5. Gill raker enlarged (×8).

FIG. 4.6. Single barb of gill raker.

that there may be a higher percentage of an organism in the stomach which was present in a lower percentage in the water.

Although zooplankton is preferred to phytoplankton, diatoms can be the most important food in the spring and autumn as judged by the weight found in the stomach. The results of a survey carried out by Hickling (1945), between 1935 and 1938, into the stomach contents of the pilchard in the Channel have been gathered into diagrammatic form. Figure 4.3 shows these results and indicates that from May through July copepods were the most important item of diet, but that from mid-February through April and again in the autumn months diatoms were most important.

Other food of the pilchard besides the three groups already mentioned, that is (1) copepods, (2) other crustacea such as amphipods, euphausids and larval forms, (3) diatoms, are (4) peridinians, (5) mollusca—especially the larvae of the lamellibranchs and gastropods, (6) eggs—mainly copepod eggs which may have been actively taken or which could have become detached from the females.

The pilchard possesses very fine structures on its gills known as gill rakers by which it is able to filter the sea water and obtain planktonic organisms which it can use as food. The arrangement of these structures is such that a perfectly efficient sieve is formed which can stop the planktonic organisms passing through it. Each gill raker consists of a backward-sloping rod on which are small barbs (Figs. 4.4 and 4.5). Under small magnification each of these small barbs appears as a fine hair, but closer inspection reveals that each is shaped on a general pattern (Fig. 4.6) and they are arranged on the main rod in the manner shown in Fig. 4.5. The angles of the single barbs to the rod vary quite considerably.

Below the gill rakers the gill filaments are arranged as shown in Fig. 4.4. These are of two lengths; the longer, curved filaments and the shorter, straight filaments. They are very vascular, and have a surface area which is greatly increased by numerous indentations.

There are five pairs of these sets of gill rakers, the fifth pair being much shorter than the other four. Similarly the fifth pair of gill filaments appeared much shorter than the others (Fig. 4.7). The way the five pairs of gill raker sets fit together to form the efficient sieve-like arrangement mentioned above is shown in Fig. 4.8. This diagram shows the appearance as seen looking into the mouth. Food caught on the gill rakers is, presumably, transferred to the oesophagus by water currents.

From the study of stomach contents through 3 years, already referred to, it was found that the pilchard has a feeding cycle which is repeated each year. The cycle shows that there are two peak feeding periods. The first comes in June and feeding, which starts about the end of February, is at a high rate between April and the end of July. From the end of July until mid-September there is a rapid decline in the amount of food taken but another peak is reached in mid-October. After this the amount rapidly falls and stays at a low level from November to February during which time practically no food at all is taken. From November to February the fish have an average of 0·1 g of food in their stomachs. Figure 4.9 is a graph made from the calculated monthly average of the weight of food in the stomach over a period of 3 years (Hickling, 1945).

Represented in this way, the data show clearly the main period of feeding with its peak, followed by a secondary but none the less important peak. This secondary peak period presumably enables the fish to survive the virtual fast which follows it until the next March.

Although the amount of food taken by individual fish falls from the end of July through August there is no decline in the amount of zooplankton available as zooplankton reaches maximum numbers in the waters south of Plymouth in August or September and copepods

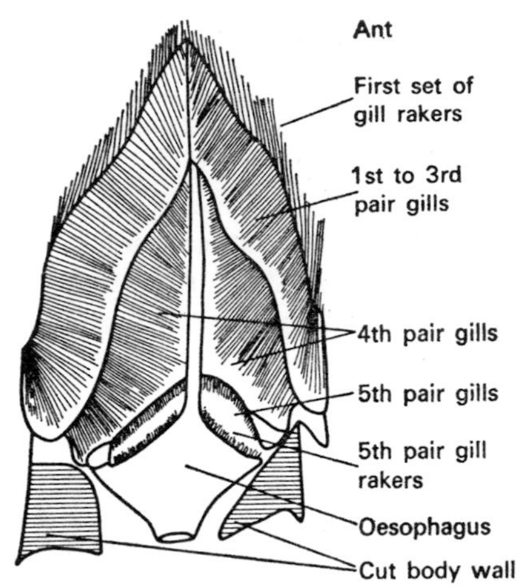

FIG. 4.7. Ventral view of gills and gill rakers ($\times \frac{3}{4}$).

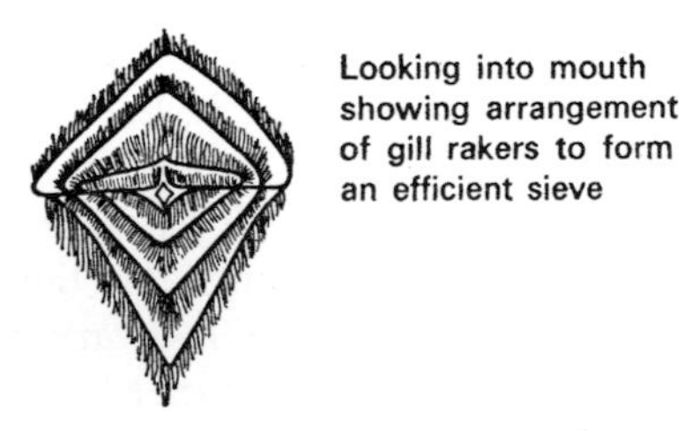

FIG. 4.8. Looking in the mouth of the pilchard showing the arrangement of the gill rakers ($\times 1$).

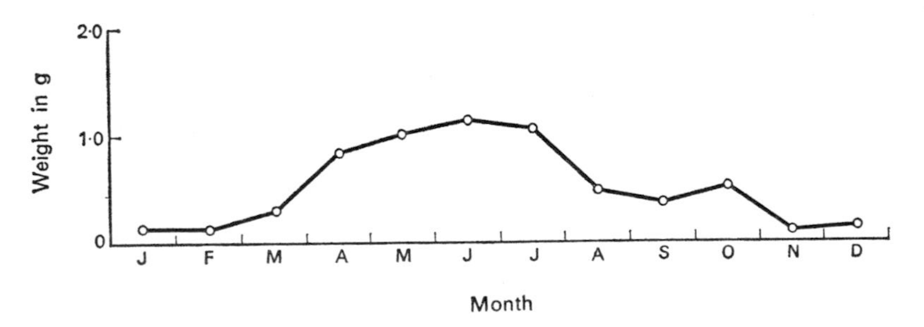

FIG. 4.9. Monthly mean weight of food in the stomach of the pilchard over 3 years.

are abundant until November. However, pilchard shoals reach their maximum abundance off the Cornish coast after June. Thus, the overall decrease in the amount of food available *per capita* of pilchard population may be the cause of the fall in the amount of food taken.

Hickling also found that the percentage of fat, in dry weight of food in the stomach, decreased during July and August and then increased again in September. Although the figures he gives are rather few, the amount of fat in wet weight of pilchards undergoes a seasonal cycle which seems largely to influence the overall seasonal weight changes. It is therefore likely that these changes in fat content of the plankton are of great importance to the fish itself. They are also economically important, as they undoubtedly affect the food value of the pilchard.

The daily feeding habits of the fish are interesting also in that there is a correlation between them and the position of the plankton in the sea. This will be dealt with more fully later. A study of the pilchard in bays revealed that at night they made a rapid migration from the shore to deeper waters. Swithinbank and Bullen (1914) suggested, rather vaguely, that this movement was due, in part at least, to an inherent instinct to find the richest feeding grounds. Zooplankton is often most abundant at the surface of the sea at night and the two movements are therefore probably correlated. Although there is apparently also some evidence that these fish feed mostly at night, it is likely that they sink lower during the hours of daylight, as the plankton do also, and continue feeding. On the other hand, it would appear from a series of observations by Lebour (1921b) that the larvae of most fish are sight feeders being unable to feed in the dark; however, this does not mean that the adult cannot continue non-selective feeding at night.

Very little is known about what constitutes the main diet of the post-larval pilchard up to about 27 mm in length. In the larval stages, which can be defined as the young fish before resorption of the yolk sac, no food at all was found. When Lebour was studying this, the smallest pilchard seen with food was 12 mm long and contained remains of a copepod nauplius. In the post-larval stages, which can be defined as the young fish after resorption of the yolk sac, from 12 to 27 mm, Lebour found no diatom or peridinian remains. However, she did find that the metamorphosed forms up to 82 mm and more, were feeding on mud which contained a large percentage of the minute diatom *Prorocentrum micans*, and that this was the most important part of the food at this stage. *P. micans* is very common and reaches its maximum abundance about September. Southward (personal communication) suggests that these metamorphosed young fish would most probably have been caught in the Tamar estuary and that their feeding habits are not, therefore, typical of this stage when found off-shore.

Fage (1920), summarizing the food requirements of the pilchard, from the small metamorphosed forms to the adult of the Channel, stated that at first their food is mainly of plant material, the animal part increasing later.

It is important to the fisherman that the preference of the adult fish for zooplankton rather than phytoplankton can delay the arrival of the fish in marketable numbers if there is a change in conditions in which phytoplankton predominates until late in the season. This happened, for example, in 1913, when the Cornish fishery was unable to work profitably until October when shoals of calanoid copepods visited Mount's Bay. Favourable conditions for the pilchard had prevailed in St. Mary's Bay in the Scillies but phytoplankton was dominant in the more usual fishing grounds off Cornwall until that date. It seems that the plankton was late that year in moving east. Thus, a knowledge of the whereabouts of the zooplankton could possibly bring with it the knowledge of the position of the pilchard shoals.

The Breeding of the Pilchard

It was once thought that the breeding grounds of the pilchard were limited to the western region of the Channel. However, since 1950 it has been known that the pilchard breeds throughout the length of the Channel although the western half is the favourite area.

It appears that during the summer there is an easterly movement of spawning fish. A line drawn from Ushant to the Scillies appears to form the boundary of spawning, judging by the numbers of eggs present. Between this line and a line drawn between Portland and

Cap Barfleur is the main spawning region. However, eggs are found east of this line and some are found in the North Sea (Cushing, 1957).

The pilchard has a very long spawning season, eggs having been found, at least in small numbers, every month of the year off Plymouth. During the 1950s the peak of the season was in June, but in recent years this has been less marked (Southward, personal communication). The gonads of both sexes, as would be expected from the date of the peak of the breeding season, reach a maximum average weight in June, although ripe testes can be found throughout most of the year, and ripe ovaries frequently from April to July (Fig. 4.10).

The pilchard is, of course, a pelagic fish, but the depth at which it spawns seems to be open to some doubt. Some authorities state that it spawns in the upper 20 m (10–11 fathoms), but Cushing (1957) found that it had a tendency to lay its eggs at midday when it was at its deepest, that is in the region of 70 m (about 35 fathoms). It is interesting to note that when working with the Californian pilchard (*Sardinops caerulea*), Ahlstrom (1943) found that the eggs were laid just before midnight in about 30 m (16 fathoms) when the fish were at, or near, their highest point in the water.

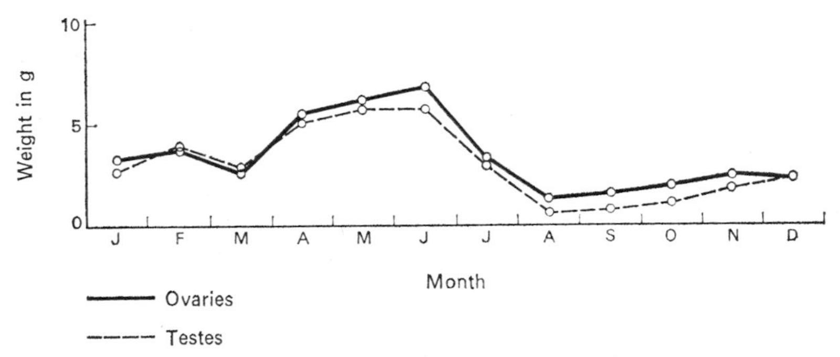

FIG. 4.10. Calculated monthly average weights of pilchard gonads.

Although it is near the northernmost limit of its range in the Channel, the English pilchard does not seem to be deterred from spawning by temperatures of 9°C in April. Average surface temperatures of the Channel (obtained from Bohnecke and Dietrich, 1951) when superimposed on some pilchard spawning maps of Cushing's (Figs. 4.11–4.14) for April to July show that where the eggs were laid in April, an area off southern Devon in which 100–300 eggs were found below a square metre, the temperature was mostly 9°C, but some may have been laid in slightly cooler water. In fact in the Pas de Calais in 1950 Cushing found some eggs in water of 8·5°C. The number at this temperature was markedly less than at 10°C. This was probably due to the smaller number of adults present, which equally well might have been a result of the lower temperature. However, 9·5°C is probably the most usual low limit for spawning, although eggs may occasionally be laid at temperatures below this.

In 1960 the White Fish Authority of Great Britain (a government-sponsored body which enigmatically also deals with the pilchard) chartered a vessel named the *Madeline* and equipped it with various types of fishing gear and electronic apparatus in order that the pilchard resources of the Channel could be studied. More will be said about this research project later, but in 1961 the research team obtained evidence which suggested that there is a general offshore shoaling and spawning in the Western Channel, and that later in the year the location is more inshore. This supported a theory which had been held for a number of

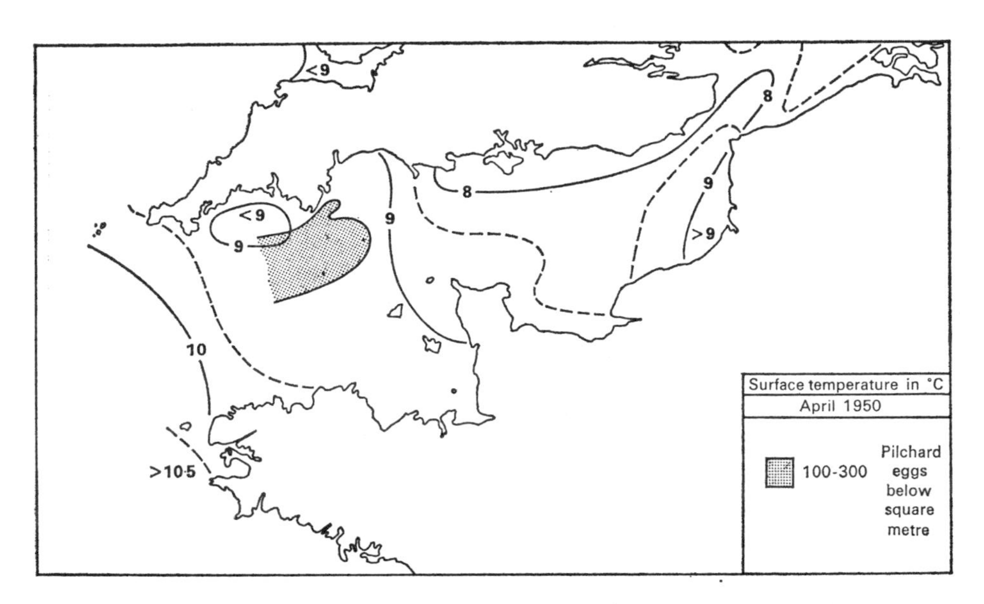

FIG. 4.11. Distribution of pilchard eggs related to surface temperatures, April 1950.

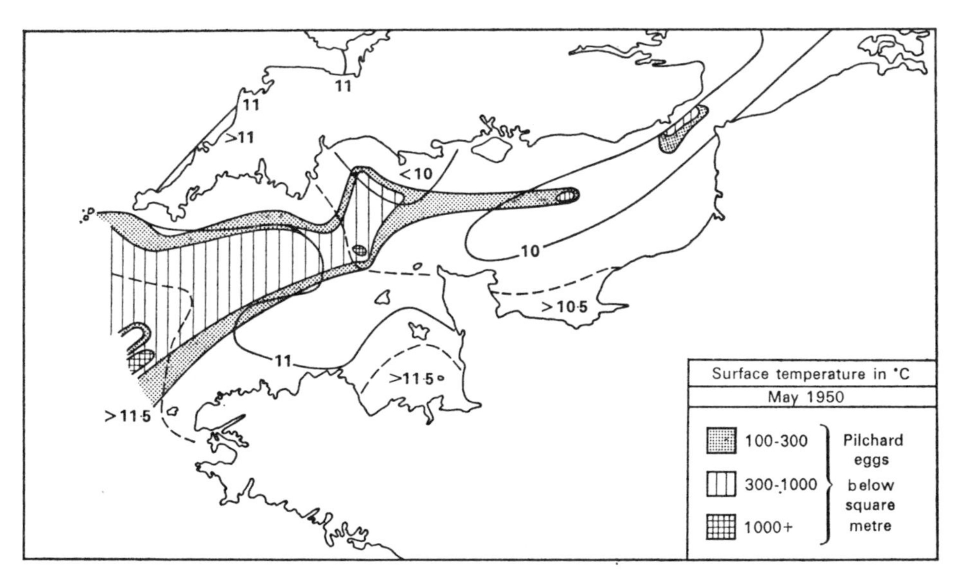

FIG. 4.12 Distribution of pilchard eggs related to surface temperatures, May 1950.

years by fishermen (Southward, 1963), and also added weight to another basic idea concerning the spawning of the pilchard; namely, that the species has a prolonged spawning period as a whole, but that it is different shoals which spawn at different times of the year.

In May 1961 Cushing (1957) reported larger numbers of eggs (300–1000 below a square metre) in the central Channel west of Portland Bill. The temperature range is mostly from 10·5 to 11·5°C during this month. However, there were two patches of eggs of more than

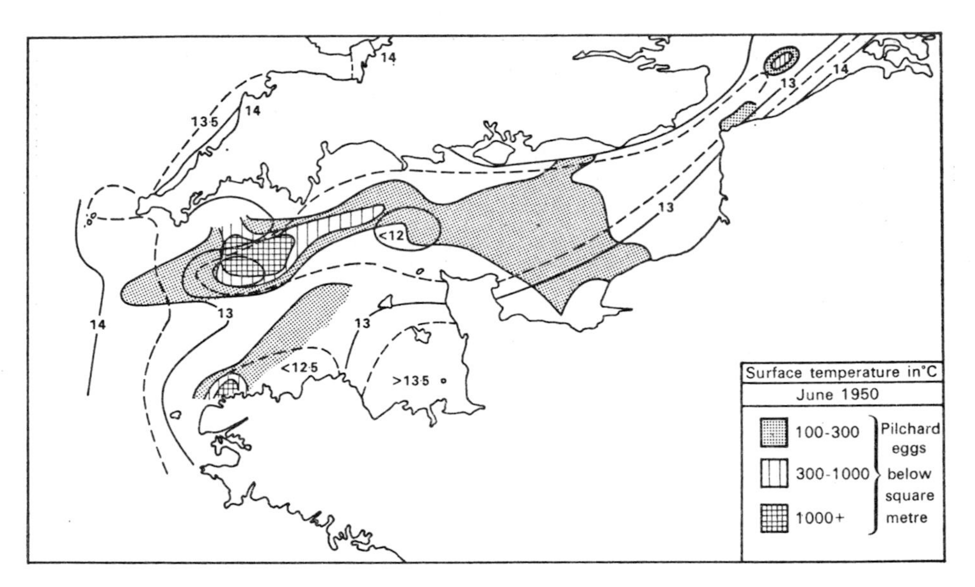

FIG. 4.13. Distribution of pilchard eggs related to surface temperatures, June 1950.

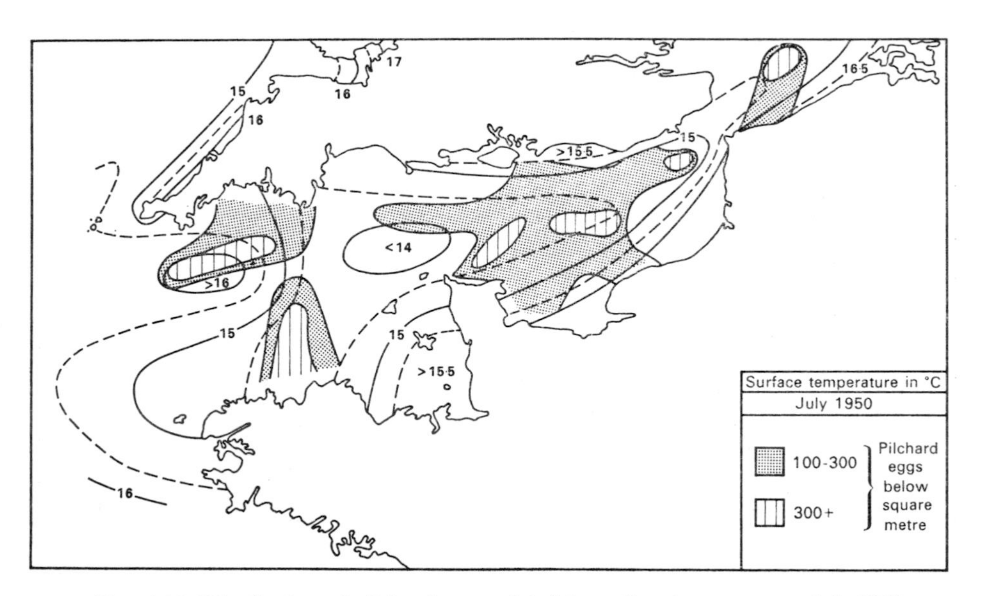

FIG. 4.14. Distribution of pilchard eggs related to surface temperatures, July 1950.

1000 below a square metre, one in water of 10°C and the other in water of more than 11·5°C. There had also been an easterly movement so that quite large numbers were found in central Channel; and there was also a fairly large patch of eggs off Dover.

When reference is made to the June chart (Fig. 4.13) a continuation of the easterly trend of movement is apparent. Eggs of density of 100–300 below a square metre were found across the width of the Channel between the Isle of Wight and the Bay of Seine. Also the

FIG. 4.15. Photographs of pilchard eggs and larvae (Plates 1–5).

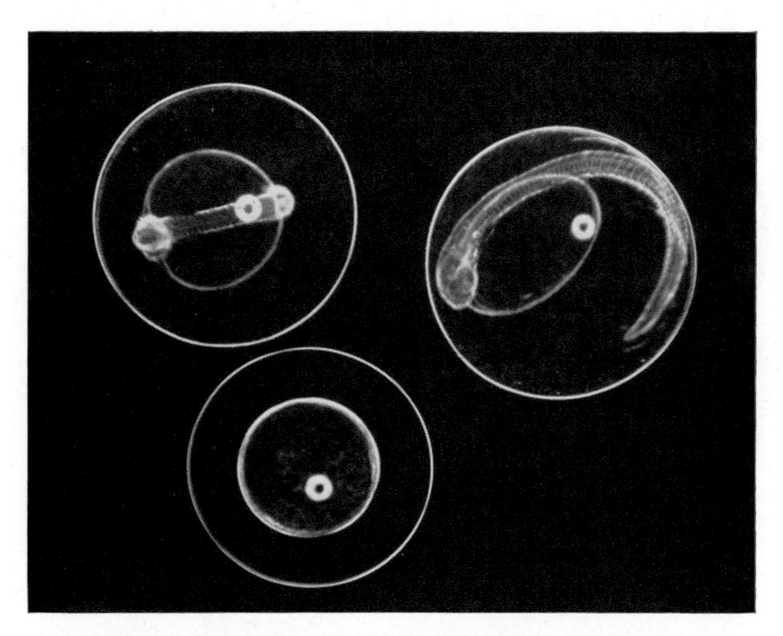

Plate 1. Eggs in three developmental stages, × 40 approx.

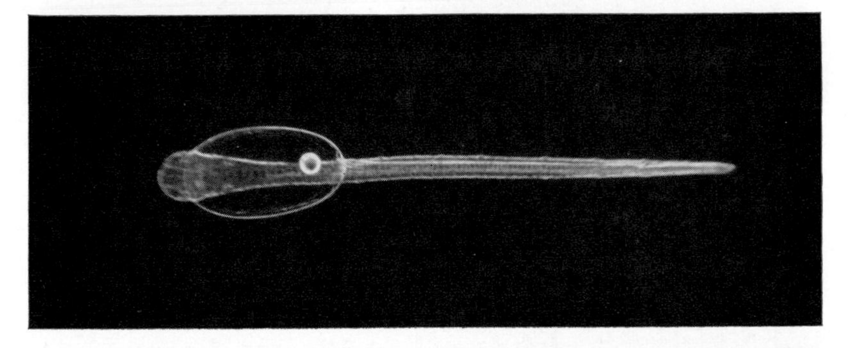

Plate 2. Larva just hatched. It floats in the surface of the water. Ventral view, ×32 approx.

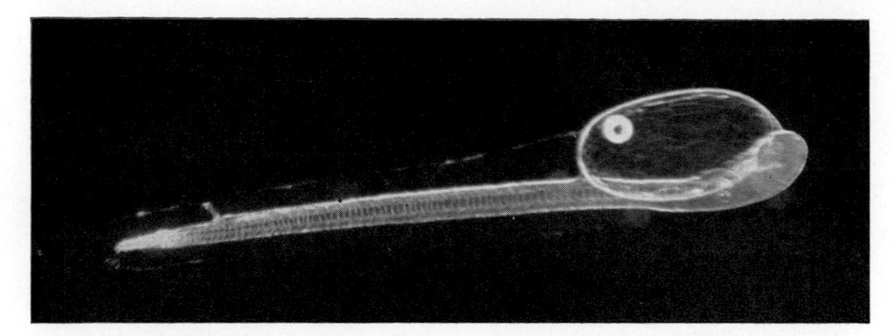

Plate 3. Larva just hatched. Lateral view, $\times 35$ approx.

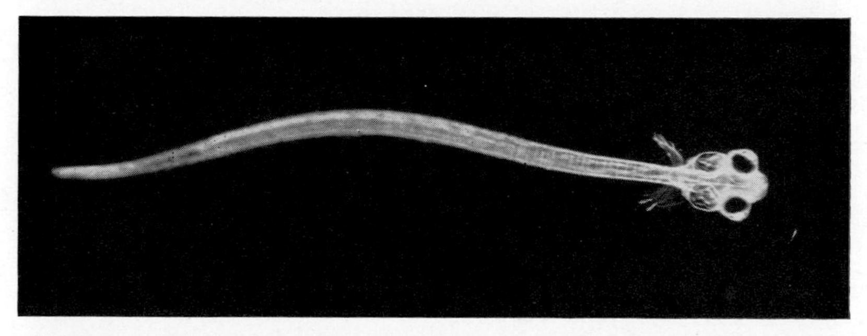

Plate 4. Larva several days old after absorption of yolk sac. Dorsal view, $\times 23$ approx.

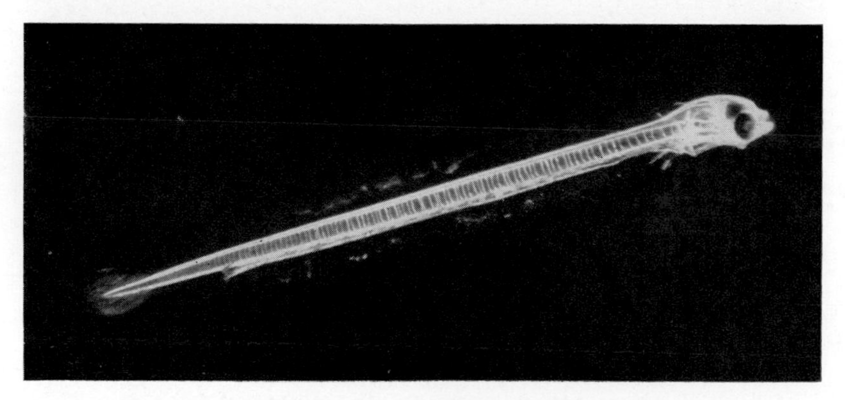

Plate 5. Larval pilchard as above but lateral view, $\times 23$ approx.

most easterly patches of eggs had been found in the Southern Bight of the North Sea. The densest patches of eggs, however, were found in the Western Channel on a line south from Plymouth. The temperature throughout this month is usually greater than 12°C.

The July chart (Fig. 4.14) shows that the largest masses of pilchard eggs had moved further east. The eastern line of this mass now extended from Dungeness, south-west to Cap d'Antifer. Most of the area of the Channel has surface temperatures of at least 14°C in July.

The information gained from a study of these four charts can be summarized as follows:

1. The intensity of breeding increases markedly in May when compared with April.
2. The most intense breeding is found in May and June.
3. A large area of the Channel is occupied by breeding pilchards in May, June and July but breeding is less intense in July than in the other two months.
4. There is a distinct easterly trend in the occurrence of pilchard eggs but the Western Channel has the most prolonged breeding season.
5. The most intense patches of breeding (found in May and June) of more than 1000 eggs below a square metre are between temperatures 10 and 13°C. In July when intensity falls off, temperatures of below 13·5°C are not found. Therefore, it may be that temperatures of 13–13·5°C are approaching the maximum favourable for intense breeding.

It is probable that more striking correlations would be found between breeding as shown by the occurrence of pilchard eggs and plankton distribution. This is further discussed in the next chapter and an example was found in the 1913 fishing season, as mentioned earlier.

It was from his survey of the pilchard eggs in the Channel that Cushing (1957) calculated the number of pilchards in the Channel. The area of the Channel from its western approaches to the southern North Sea was divided into rectangles each of which contained sampling stations. From these he was able to obtain an average number of eggs for each rectangle and thence for the whole of the area considered. By dividing this figure by the number of eggs produced by an average-sized mature female he determined the approximate number of mature females in the stock. The sex ratio of the population as a whole was then taken to be unity thus giving an estimate of the number of adult pilchards in the Channel at that time. It was from these figures that Cushing then estimated roughly the wet weight of the Channel pilchard stock.

Life History

A female pilchard lays between 60,000 and 80,000 eggs during one breeding season. These eggs can be identified by the three following characters:

1. Possession of an unusually large peri-vitelline space (Fig. 4.15, plate 1).
2. A single, large oil globule in the vitellus (Fig. 4.15, plate 1).
3. A completely subdivided yolk.

The rate of development of the egg to hatching is variable depending on the temperature of the surrounding water. At 16.4°C the egg is 3 days before the larva hatches from the egg. The larva is about 3·8 mm long on hatching. It possesses a yolk sac and a sensory papilla on the ventral surface (Fig. 4.15, plates 2 and 3). Ahlstrom (1960), working with the eggs of the Pacific pilchard, showed that at 12°C eggs took 4 days to develop from fertilization to hatching, at 14°C and 16°C the times were 3 days and 2 days respectively. Cunningham (1891) found that at 17°C the eggs of the Cornish pilchard developed to hatching in 3 days. When newly hatched, besides the large papilla there are also protuberances on each side of the body. These are larval sense organs from which, in fish which possess one, the lateral line system later develops. However, in the pilchard, which lacks a lateral line, these must be lost in the course of development.

When first hatched the yolk is large and the mouth has not opened. On the third day after hatching, the mouth opens on the ventral surface of the head. Also, at this time, some pigment begins to appear in the body and the eye. Figure 4.16 shows a 9-day-old pilchard and Fig. 4.15, plates 4 and 5 are photographs of a similar stage of development.

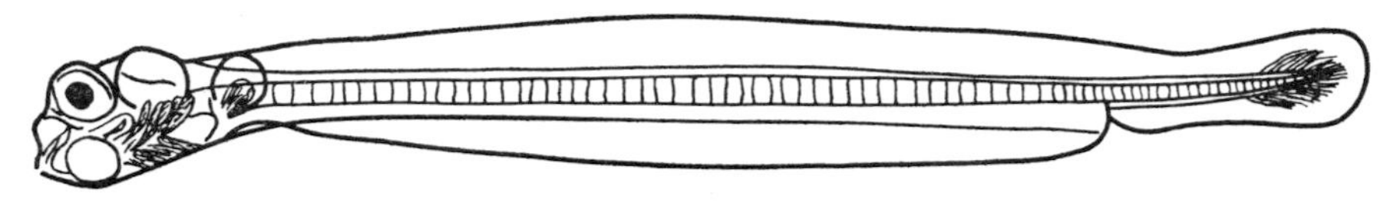

9 day old Larva of Pilchard (After Cunningham)

FIG. 4.16. Nine-day-old pilchard larva.

The eyes and gills are quite clear and the auditory ossicle is shown, but no mouth has yet developed. By the tenth or eleventh day after hatching, the yolk in the yolk sac is exhausted, and the mouth, with jaws developing, has moved to a terminal position. The larvae begin to take food for themselves. On the twelfth day they have grown to about 5·5 mm in length.

After about a year, although this depends on the time of spawning, the young reach a length of from 10 to 15 cm (Cunningham, 1891). This rapid rate of growth continues until, by the end of the third year the fish have attained a length of about 19 cm. After this time, however, growth slows very considerably being in the order of 1 cm or less per year. The maximum length recorded is in the region of 27–28 cm.

Usually by their fourth year the fish have become sexually mature although, again, the time of attainment of maturity is largely dependent on the time of year in which the individual was spawned. These fish, whose growth rate has slowed and which have attained sexual maturity, are now ready to start the life cycle again. The females lay their numerous eggs which are fertilized externally by the males. The eggs are planktonic and great wastage of both these and of the hatched larvae occurs.

CHAPTER 5

DISTRIBUTION AND MIGRATORY MOVEMENTS

Distribution

The distribution of the Cornish pilchard in the Channel will be considered in relation to four environmental factors. These are thought to influence the movement of fish shoals and are:

1. water temperature,
2. plankton distribution,
3. currents, as interpreted from the information gained in 2,
4. light.

The distribution of pilchard eggs throughout the Channel has already been shown (Figs. 4.11–4.14). On the areas where the eggs were found by Cushing (1957), isotherms were superimposed and it was decided that no definite conclusions could be drawn from the relationship of eggs to temperature except perhaps that pilchards were not found in any abundance below 9·5°C. The small numbers of eggs found in water of 9°C and 8·5°C may have been due to a smaller number of adult fish in such cold water.

It is frequently stated that the pilchard is near the northern limit of its range in the Channel. From evidence on migration (see later) it may be truer to say that there is a population of pilchards which stays in the area of the Channel and the eastern Atlantic. Here they have become adapted to the temperature range, which is more extreme than further south. However, they are thought to move to the deeper water of the eastern Atlantic at the edge of the continental shelf, during the winter, which is warmer than the water of the Channel.

1. *Phytoplankton*

Although it has already been pointed out that the pilchard eats mostly zooplankton there is some statistical correlation between pilchard abundance and intensity of phytoplankton in the Western Channel in May and June. This correlation can be shown also over the whole area of the Channel. Cushing (1957) came to the conclusion that the aggregations consisted of non-spawning fish; either those which are not yet mature enough to spawn, or those which are spent and drifting more or less passively, recovering and feeding on the phytoplankton. Numbers of spawning fish were not shown to occur on phytoplankton.

2. *Zooplankton*

During the cruises of the R.V. *Sir Lancelot* in 1950 in which the work by Cushing (1957), already referred to, was done, an attempt was made to ascertain whether there was any particular plankton animal with which the pilchard appeared to be associated. No positive results were obtained, but when zooplankton of certain sizes were grouped together, slight positive correlations were found.

The zooplankton was divided into four size-groups as shown in Table 3.

TABLE 3. ZOOPLANKTON FROM THE CHANNEL CLASSIFIED INTO SIZE-GROUPS (after Cushing, 1957)

Size-group	I	II	III	IV
Size	1 mm³	1–2·5 mm³	2·5–5 mm³	5 mm³
Animal	*Calanus* *Paracalanus*	*Calanus* *Paracalanus* *Pseudocalanus* *Acartia* *Centropages*	*Temora* *Pseudocalanus*	*Calanus*

With the zooplankton classified in this way Cushing found that early in the year, in April, there was a marked correlation between pilchards and size-group I. Owing to the western position of the fish this was only found in the Western Channel during April, but when the fish had spread to the eastern part of the Channel also, a slight correlation with all size-groups was shown.

The shoals of fish were recorded with an echo-sounder, and maps of the distribution of fish soundings in the Channel for April to July 1950 are shown in Figs. 5.1, 5.3, 5.5 and 5.6. These are considered to be mostly pilchard but small quantities of other pelagic fish may be included. Maps showing traces in the Western Channel only, during March and April 1961 (Fig. 5.2) and May 1961 (Fig. 5.4) are also shown. (These were kindly given to me by Mr. J. P. Bridger, who was in charge of the White Fish Authority research programme in the area, already referred to.) In comparing the maps for the two sample years, three points of interest emerge. In 1961, first the fish were more westward during April; secondly, during May the

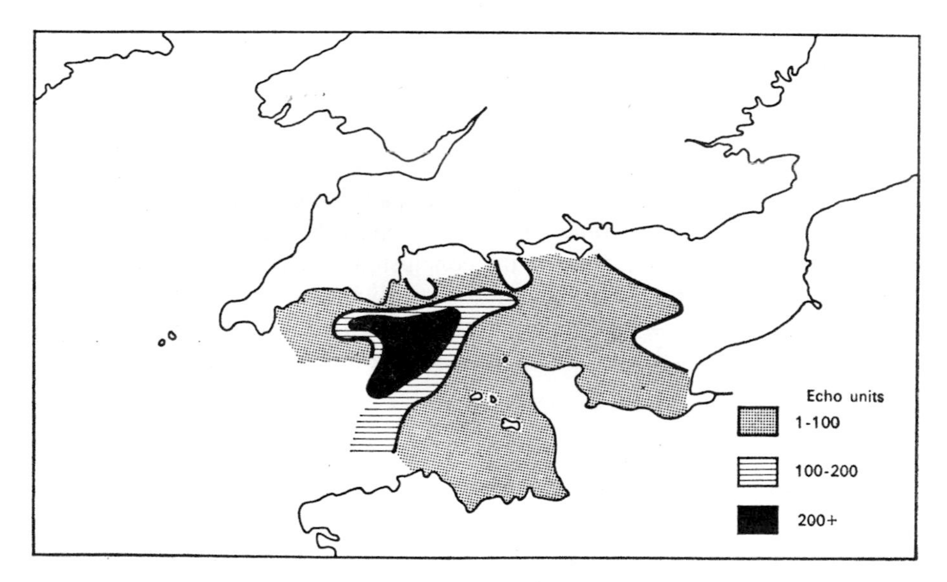

FIG. 5.1. Distribution of pilchards in the English Channel shown by echo traces, April 1950.

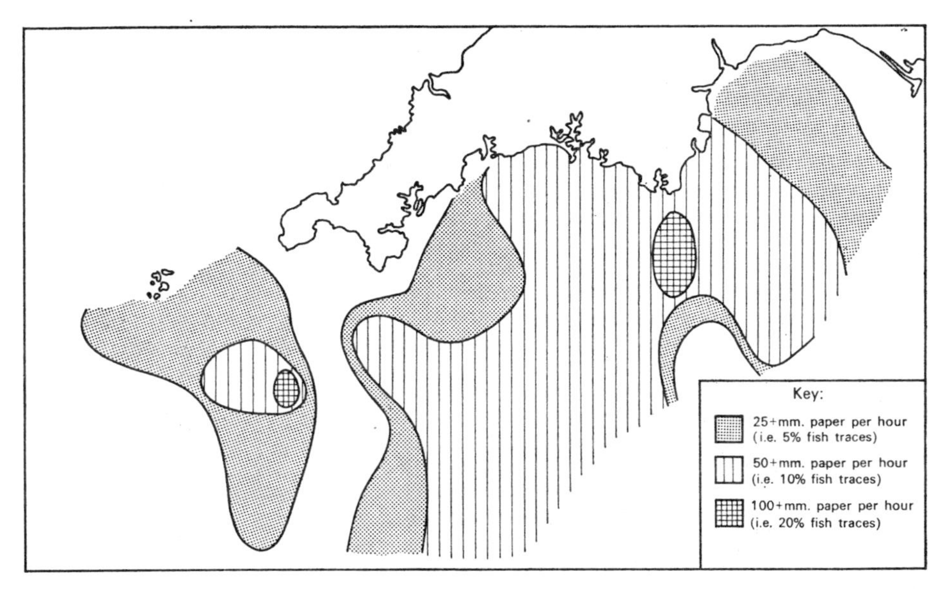

FIG. 5.2. Distribution of pilchards in the English Channel shown by echo traces, March to April 1961.

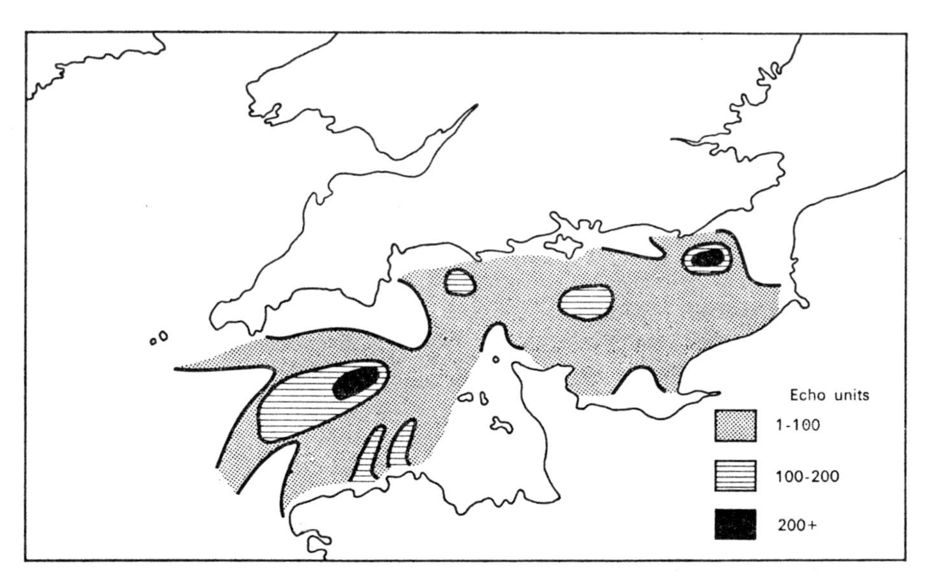

FIG. 5.3. Distribution of pilchards in the English Channel shown by echo traces, May 1950.

main body of fish had also remained west; and thirdly, there were very much fewer traces in the northern regions of the Western Channel than for May 1950.

The echo-sounder used by Bridger on the R.V. *Madeline* used a maximum of 500 mm of paper per hour. Thus, a score of 500 meant that echos were being returned continuously, representing a solid body of fish; 100 mm/h meant that fish traces were found 20 per cent of the time and 50 mm/h meant that fish traces were recorded for 10 per cent of the time. The information derived from these echo-soundings in the Channel has been used to develop

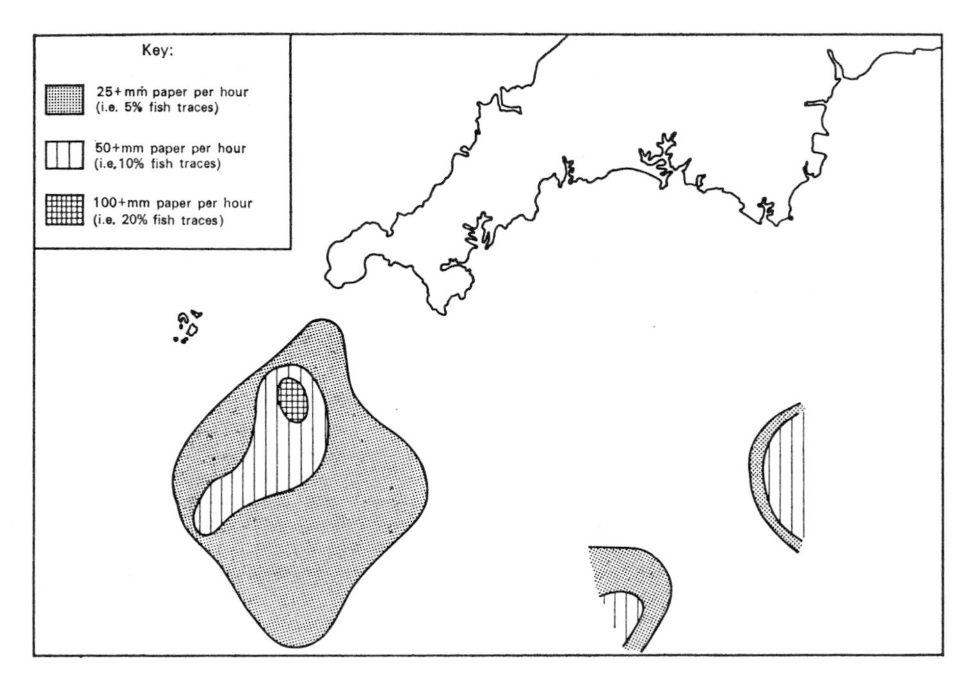

FIG. 5.4. Distribution of pilchards in the English Channel shown by echo traces, May 1961.

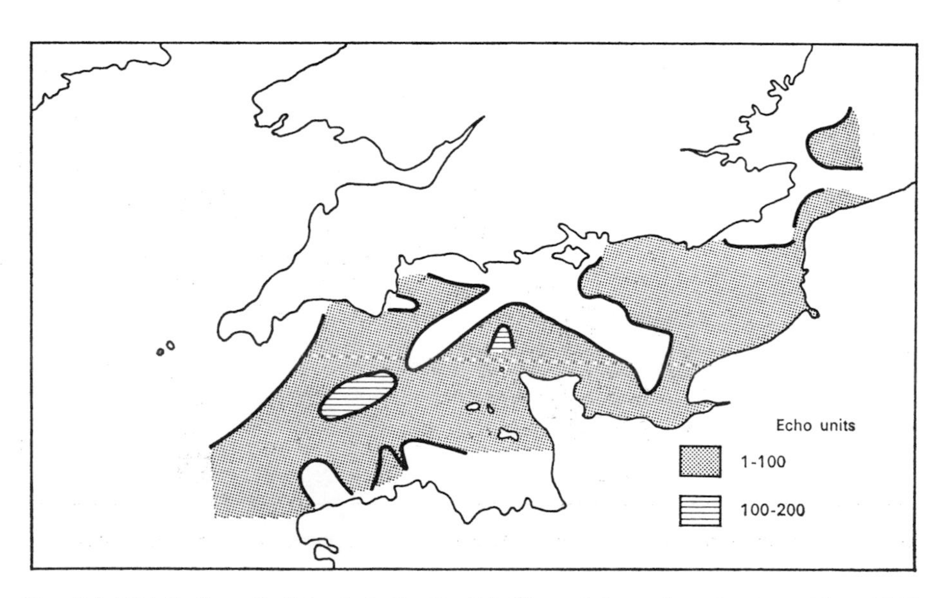

FIG. 5.5. Distribution of pilchards in the English Channel shown by echo traces, June 1950.

a migration hypothesis which is suggested under the section dealing with some migratory theories which have been proposed for the pilchard.

The zooplankton found during 1950 will be considered in more detail as this information gives an indication of the way in which plankton and pilchard movements may coincide (Cushing, 1957). First, there was a patch of *Calanus finmarchicus* off Start Point in April

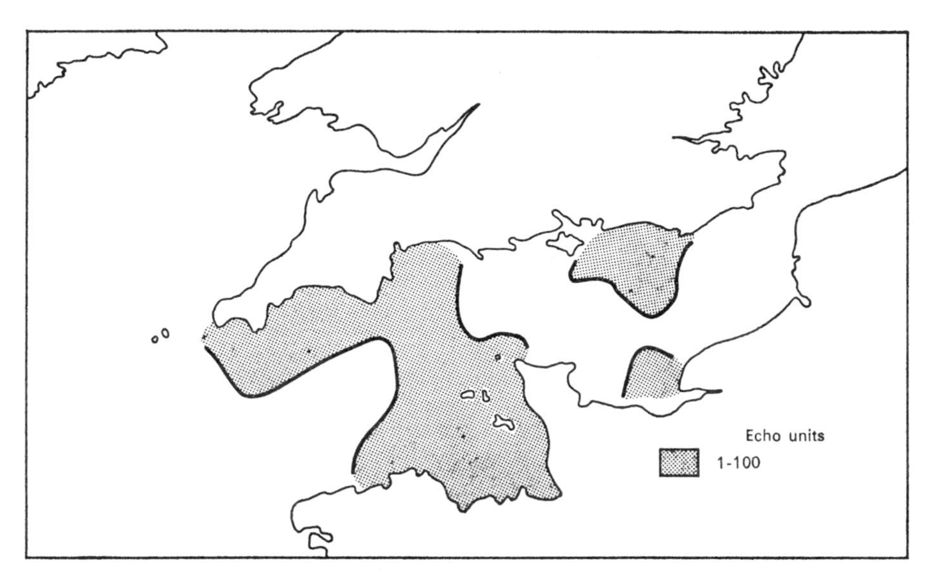

FIG. 5.6. Distribution of pilchards in the English Channel shown by echo traces, July 1950.

which, from April until July, gradually moved south. This was later followed by another patch which moved up from the south-west. A similar path was followed by the patches of *Pseudocalanus elongatus* and *Paracalanus parvus*. During this time, after an early correlation between pilchards and *Calanus*, some of the fish broke away from the plankton and moved into the eastern Channel. Others apparently tended to move south from Start Point with the *Calanus* or *Pseudocalanus* patches.

Certain zooplankton species tended to be spread more generally over the Channel and remain rather static in their distribution. *Temora longicornis*, *Acartia clausi*, *Centropages typicus* and *C. hamatus* are examples of this type. The two *Centropages* species particularly showed a static distribution, *C. typicus* being confined to the Western Channel and *C. hamatus* to the Eastern Channel as shown in Fig. 5.7. The two species were not found to mix; *Calanus* existed side by side with *Centropages typicus*, and *Temora* with *C. hamatus*. It was thought that in July the distribution of the fish may have been related to that of *Temora*.

3. *Currents*

It appears that during 1950 there was very little overall movement of the Channel water. In the Western Channel there appears to have been a clockwise movement of water moving from the south-west across the Lizard to Start Point and then moving south. As was pointed out above, some of the pilchards followed this movement. Others swam further east and to do so did not follow any current movement.

The line dividing the two *Centropages* species runs approximately from the Isle of Wight to the Gulf of St. Malo passing to the west of the Channel Islands. This line is approximately also the line which divides the Channel on other characteristics. For example, west of the line the waters are less turbulent, as the tidal streams are weaker than to the east, which has strong tides. Also, due to the tidal effect is the fact that in the west there is sometimes a thermocline during the summer, but this is not so to the east of the line (Dietrich, 1951; Holme, 1961, 1966).

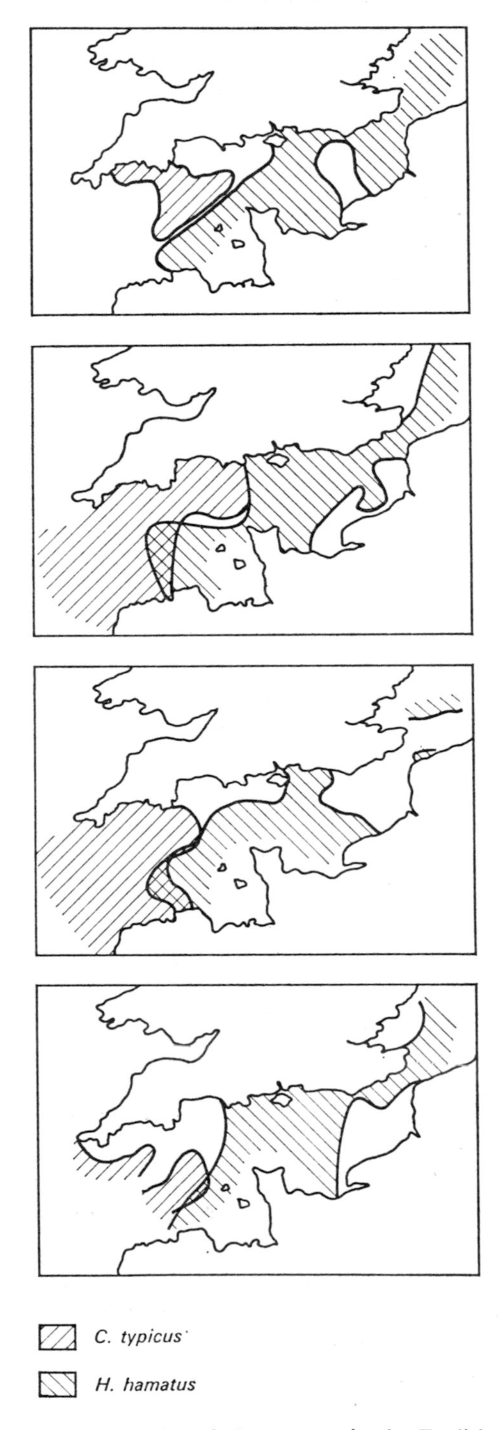

FIG. 5.7. Distribution of two species of *Centropages* in the English Channel, 1950.

Thus, the *Calanus*, *Pseudocalanus*, *Paracalanus* and *Temora*, which were in general found in the western region of the Channel, tended to remain there, and the *Acartia* stayed mostly in the east although some were found also in the west.

It would seem, therefore that the western region of the Channel was an area which, at least in 1950, was the most heavily populated with zooplankton, thus encouraging the pilchard to remain there. Those which moved east could, however, graze on *Acartia* and *Centropages hamatus*.

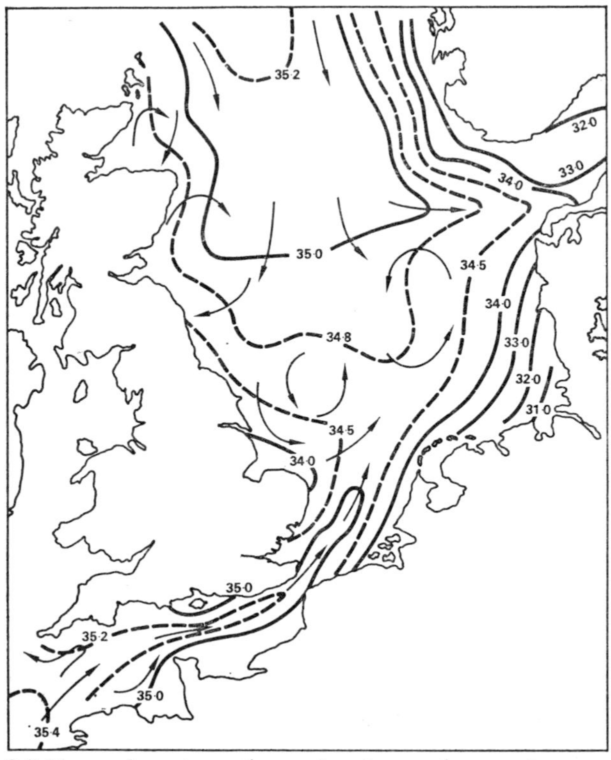

Salinities are in parts per thousand; a tongue of more saline Atlantic water projects into the mouth of the Channel

——→ Water circulation

FIG. 5.8. Typical salinities around the British Isles in winter.

The currents of the English Channel will now be dealt with in a little more detail, especially in relation to certain plankton species. As mentioned above, the waters of the Channel tend to remain as a rather separate body of coastal water, less saline than the oceanic waters of the Atlantic, which often have a salinity of 35·4–35·0‰. The salinity of mid-Channel waters is a result of an admixture of oceanic waters of higher salinity and coastal waters with a salinity of about 35·0‰. The lower salinity in coastal regions is due to the addition of fresh water from rivers. A map showing typical salinity values in the seas around the British Isles is shown in Fig. 5.8. The tongue of water with a salinity value of 35·4‰ is an incursion of Atlantic water in the mouth of the Channel, and indicates the short eastward distance which it tends to travel along the Channel (Lumby, 1935). (This distinct character of the salinity of various waters has been used extensively in charting ocean currents of the world.) Another characteristic of Channel water is that it is relatively poor in nutrient salts and its plankton population is poorer than, and different from, that of other seas surrounding the British Isles.

Agassiz (1883) was probably the first to point out that certain animals, when found in water in one place, could be said to indicate that the water had travelled from another place. Later, this idea was greatly developed by Russell (1935, 1936, 1939) and others. Particular animals are often peculiar to one type of water, presumably because conditions within that body of water are more suitable for their survival than elsewhere. These animals are called plankton indicators. For example, *Sagitta setosa* is a species of arrow-worm frequently found in the Channel but not in other waters around the British Isles. Its presence in a sample of water suggests that there will be a general paucity of species though not necessarily a paucity in numbers of individuals.

Some years ago it was thought that the extent of the incursion of oceanic waters into the Channel had lessened. Before 1930 water off Plymouth used to contain a higher content of nutrient salts in winter, and was characterized by the presence of *S. elegans*. East of Plymouth was more coastal water containing *S. setosa*. Since 1930, however, the boundary between these two types of water has been rather further west so that the area off Plymouth has been characterized by *S. setosa* water. The change appeared to be very abrupt because conditions at the Plymouth stations showed a marked difference between 1930 and 1931.

However, Southward (1962, 1963) suggested that the changes in the Channel have been due to a gradual warming of the water. The mean surface temperature near Plymouth from 1928 to 1959 was 0·5°C higher than the mean from 1903 through 1927. The apparent abruptness of the change off Plymouth can be explained by postulating an advance in the boundary between the warmer and cooler water effected by a slight change of water circulation. Thus, an emphasis of water from the south-west on the circulation of the Western Channel would introduce more eastern water into the circulation making *Sagitta setosa*, which is dominant in the eastern Channel, become suddenly dominant at certain stations in the Western Channel.

Nevertheless, the change in fauna actually sampled was abrupt for it was found that in 1939 94 per cent of the *Sagitta* species present off Plymouth were of *S. elegans* and 6 per cent *S. setosa*. In 1931, however, *S. setosa* constituted 83 per cent and *S. elegans* 17 per cent of the *Sagitta* population. In under 10 years from this change, the herring fishery of Plymouth which used to yield sometimes 80,000 cwt p.a., had dropped to under 30 cwt. Although this was catastrophic for the herring fisherman, it is possible that this change has helped the pilchard population in a way that will be mentioned later.

A map showing the distribution of the *Sagitta* indicator species around our coasts is given in Fig. 5.9. The map shows conditions in which there was a large influx of water from the Atlantic into the North Sea from the north. If this current is very strong it may hold up an overflow of water from the Channel to the North Sea via the Straits of Dover. The map also mentions another *Sagitta* species; this is *S. serratodentata*, which is typical of the open oceans. It may occur in company with *S. elegans*, depending on how much the open water has mixed with the coastal water.

Another feature of interest concerning the currents in the mouth of the Channel was discovered by Matthews in 1911. He found that a counter-clockwise cyclonic circulation existed there. In winter the circulation persists but has its boundary further north. This phenomenon has been described by Harvey (1925, 1930, 1950), and Lumby (1925) stated that the southern limit of this circulation is approximately in the latitude of the Scilly Isles in winter, but in summer it extends as far south as Ushant. The result of this southward extension of the boundary is that the mouth of the Channel becomes effectively blocked to inward (that is, eastward) movements of water except from the south west. This is borne out

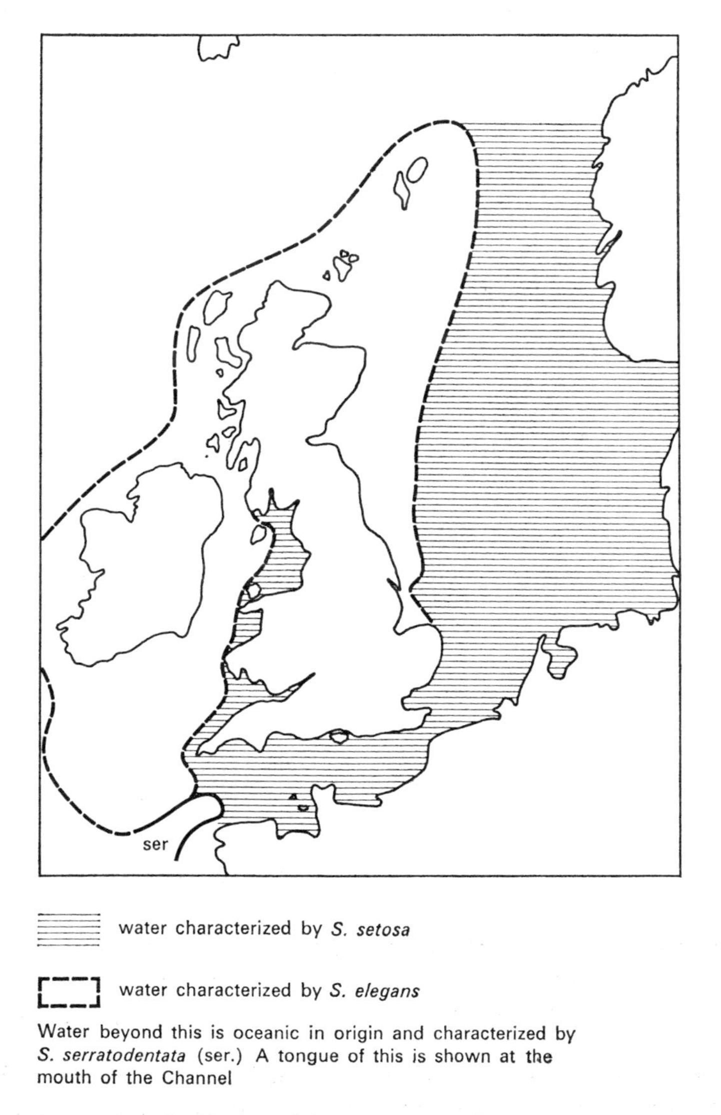

FIG. 5.9. Typical distribution of three *Sagitta* species around the British Isles.

by the observations on plankton which suggest that after May no influx into the Channel of coastal water, with which *S. elegans* is associated, occurs. However, as shown in Fig. 5.8, a tongue of water from the south-west at least reaches the mouth of the Channel as indicated by the presence of *S. serratodentata*.

It is interesting to note here that Russell (1939) stated that for a number of years the water in the Western Channel has been remarkable for its "lack of plankton indicators including both *Sagitta* species". However, it had "been characterized by large numbers of pilchard eggs" and had been called "Pilchard Water".

Hardy (1956) noted that these different waters are visible from the air by colour differences, on this particular occasion the "*setosa* water" was greenish brown or green and the "*elegans* water" blue; but "*elegans* water" can be green and south-western (oceanic) water is often blue (Southward, personal communication).

4. *Light*

Although for the sake of convenience these four factors have been considered separately, it is really impossible to isolate conditions from one another in this way, especially when dealing with subjects such as oceanography and hydrography. For example, it was found when dealing with currents that these were affected by temperature, that they affected

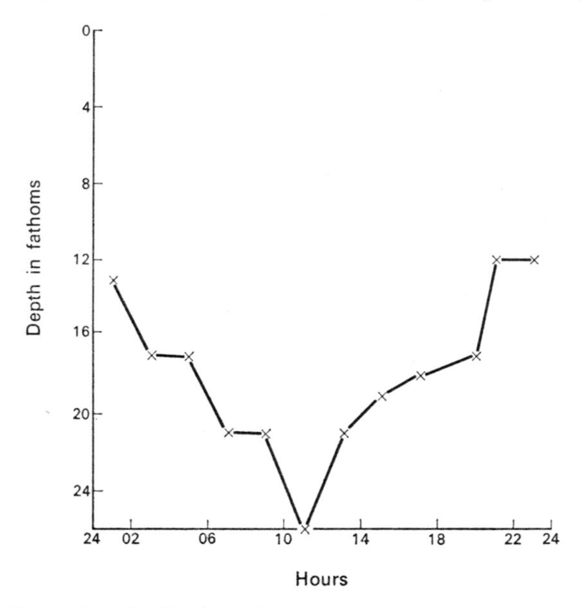

FIG. 5.10. Vertical distribution of *Sardina pilchardus* over 24-hour period.

plankton distribution, which in turn governed the movements of fish. Now, in dealing with light, it is found that this affects plankton distribution and therefore, to a large extent, will change the position of the aggregations of fish feeding on, or spawning near, the plankton patches.

If a net is being used to catch plankton, it is found that the catch at night tends to be large for most species. On what might be termed a "bright" day the catch at the surface tends to be good, but if the light intensity is really great the catch is poorer. Thus, plankton tends to move towards a light source until it reaches a certain maximum; above this it swims away from it. A complicating factor here is the ability of plankton to avoid the net during the day time. It is often found that on a day when the light intensity is great, the maximum catch is made at a depth of from 5 to 10 fathoms; on a moonless night the plankton is scattered randomly. In the Arctic region plankton are near the surface both day and night. Each plankton species has its own maximum light intensity toleration.

Calanus may be taken as an example of a plankton genus which shows typical diurnal movements related to light intensity. At dusk *Calanus* move towards the surface and are found collecting at the surface within the top metre. At nightfall their distribution becomes random, but at dawn there is again a tendency to aggregate at the surface. With increasing light intensity they sink lower in the water, and by noon they are several fathoms down. As the light intensity diminishes they rise again in the water until they reach the surface at about dusk. However, this pattern of movement is found in other genera and a review of the vertical distribution of zooplankton can be found in Raymont (1963, pp. 418–66).

The vertical distribution of the pilchard was plotted on three cruises of the *Sir Lancelot* in 1950, and the mean of these depths is shown in Fig. 5.10. It can be seen that they reached their greatest depth (26 fathoms) just before noon and gradually rose in the water later in the day until at midnight they were only 12 fathoms down. At day break they began to fall again and by 7 a.m. they had reached a depth somewhat greater than 20 fathoms. It is possible that this pattern of distribution, taken in the months of April, May and June, might alter slightly during other months when the light intensities are different.

Migration

Very little is known of the pilchard's migratory movements. For example, there is no evidence as to what happens to the adults in the winter and the young fish before they reach 3 years of age. It should be noted that the movements of pilchards that have been described for the Channel area are based on circumstantial evidence as no tagging experiments have been done, and this should be borne in mind when the following paragraphs are read.

For adult pilchards the available data suggest an approach to the Western Channel in April, and the spawning charts already referred to (Figs. 4.11–4.14) show the areas throughout the Channel in which pilchard eggs are found and to which, therefore, adult pilchards have probably travelled, although currents may have changed their original location to some extent. The impression gained from studying these charts is one of gradual easterly migration possibly by one main body of fish, but the facts now known suggest that different shoals of fish may be concerned. It seems that one body of fish may lay its eggs in one area, and later another body of fish, having moved further east, lays its eggs. This pattern of events was noted in 1947 when some experiments, organized by the Ministry of Agriculture and Fisheries, were carried out. Some large pilchard shoals were located off Plymouth before other shoals had left St. Austell Bay. When those fish in the bay did move they did so in a southerly direction rather than moving east towards Plymouth (Hodgson and Richardson, 1949).

Norman (1931) stated that the pilchard "always retires to warmer regions on the approach of winter". However, those fish that have moved east along the Channel as far as the southern North Sea and spawned, are believed to make their way south-west when they are spent and are found off Boulogne in autumn and early winter. They then appear to continue westwards towards the main body, but not south as Norman implied.

Kennea (1957) put forward a hypothesis suggesting that the pilchard moves westwards and over-winters in the western approaches of the Channel. This theory was proposed as pilchards are occasionally trawled from the sea bed during winter along the edge of the continental shelf in the eastern Atlantic. It is closely parallel to the theory put forward some years earlier concerning mackerel migration in the area (Steven, 1948). That the adults first appear off Wolf Rock in March, and the first pilchard eggs are found in April in the south-west Channel off southern Devon, lends further weight to this argument.

It seems that the pilchard population of the Channel and extreme east Atlantic is probably largely separate from other pilchard populations, although there may be some influx from the area off Portugal during the north-easterly trend of movement in the spring. It is probable that the young of all groups of pilchard in the region follow the adults to the south coast of the Channel where they grow to sardines and stay till they are 2–3 years old. They then move north and east again in the spawning cycle. Recent information suggests that the pilchard is again altering its habits as the sardines formerly found in large numbers off the north coast of France are disappearing (Bridger, personal communication).

Meanwhile the adults leave Cornish waters in December and move to the edge of the continental shelf in the eastern Atlantic and stay on or near the sea bed in deep water, until March or April when they, too, move east and slightly north again.

Probably the necessity of maintaining a food supply is the factor governing the paths of migration followed by the pilchard. It seems that spawning fish aggregate on zooplankton following the *Calanus* or *Pseudocalanus* which often drifts south from Start Point. If phytoplankton is present Cushing says that non-spawning fish may aggregate on it, as discussed previously in detail. The temperature also probably affects migration to some extent mainly in the effect it has on the production of plankton.

Cushing (1957) also suggested that the migratory movements might be linked with the hydrographic phenomena of the Channel mentioned earlier; namely the line from the Isle of Wight to the Gulf of St. Malo which divides an area of turbulent water and no thermocline to the east, from less turbulent water associated with the presence of an occasional summer thermocline to the west. Because there is considerable evidence that most of the population stayed in the west he proposed that the fish merely follow the water movements in the area which are in an approximately clockwise circulation. However, in autumn, according to Dietrich (1951), the direction of circulation reverses, thus bringing the fish back to Cornwall.

This theory, however, does not account for the eastward movement of at least part of the population. Once the fish have broken away from the cyclonic circulation in the Western Channel, and passed the "dividing line", currents which are known to flow through the Straits of Dover take the fish with them. These may finally end up in the southern North Sea and as we have seen, after spawning there they return in the autumn to the area off Boulogne.

The recent work of Southward (1963) emphasized what has already been suggested, that there are two groups of pilchards in the Western Channel. These he called a "western" group and a "south-western" group. He suggested that it is the "western" group which spawns along the British side of the Channel, but offshore. The "south-western" group is probably that which spawns inshore along the Devon and Cornish coasts during the last 3 or 4 months of the year.

Even if it is postulated from this evidence that two races of pilchard exist with different spawning and migrating habits, it still does not account for the movements of the fish in the eastern Channel; and, although Southward said that there is no evidence of shoals returning from the Eastern to the Western Channel, he did not offer an alternative explanation for the proven presence of pilchards in the Eastern Channel. On this subject he said that "the apparent movement of shoaling and spawning" from west to east in the Channel is probably due to "differences in the habits of fish resident in the various parts of the Channel".

This implies yet another group of pilchards which could lead to the hypothesis of three races in the Channel.

However, returning to the "western" and "south-western" groups of pilchard, Southward suggests that the latter has probably originated in the Biscay region, is adapted to warmer, south-western water, and is "certainly the group of the traditional drift and seine net fishery, but, from the relative abundance of it, would appear to be much inferior in numbers of adults to the "western" group. In his paper Southward had presented evidence of a considerable increase in the stock of pilchards in the Channel since the 1930s. He suggested that it was in the "western" group that this had occurred and that it was largely attributable to the gradual rise in water temperature of the Channel since about 1930. Even small temperature increases would be advantageous to a basically warm-water species in a cool environment, and they could be correspondingly disadvantageous to a cold-water species at the southern limit of its range, such as the herring. Most important of all from the industry's point of view was the suggestion that this "western" group has hardly been fished by the traditional pilchard fishery, the "south-western" group forming the basic stock of the industry at present.

Apart from the main migratory movements, incidental movements also occur in that, after they have spawned, pilchards tend to move shorewards before continuing on their migratory paths. Later in the year, from July until the autumn, the shoals disperse and are difficult to locate on echo-sounders. It is this type of behaviour which determines what gear can best be used for their capture.

CHAPTER 6

HISTORICAL PERSPECTIVES

NO INDUSTRY can be studied effectively without knowing something of its history as the reasons for its present structure are inevitably based in the past. In the next few chapters it will be seen that although for many years the pilchard industry remained more or less unchanged in its most important aspects, in the past 50 years there have been rapid changes in techniques of catching and preserving the fish. Both these activities are fundamental to the continuing life of the industry and will be dealt with in detail in Chapters 8 and 9 respectively.

Detailed records concerning the industry are not available before the middle of the eighteenth century. Previous to this time the pursuits of Cornish fishermen were occasionally referred to in historical documents, but these occasions were usually when clashes with the authorities occurred. The incompleteness of the picture that can be obtained of the industry in the past can probably be mainly ascribed to its geographical isolation.

The first important references to the pilchard industry seem to be those dealing with the Salt Laws which were imposed and lifted several times over a period of 131 years from 1694. The pilchard industry is mentioned in this connection because the export of salted or cured pilchards to Italy was considerable as far back as the sixteenth century. At the same time that salt duties were imposed, a bounty of 12*s*. for every 50 gal hogshead of pilchards exported was introduced. The actual duty on its first imposition was 1½*d*./gal on English salt and 3*d*./gal on imported salt. In 1718 the bounty was reduced to 7*s*. per 50 gal hogshead (Rowe, 1953).

By the middle of the eighteenth century (for example, during the 10-year period from 1747 to 1756) the average annual weight of exported cured pilchards was 29,795 hogsheads. As a hogshead weighed about 4 cwt gross, this is about equivalent to 6000 tons. The annual value at this time of these exports, including the bounty, was estimated at £50,000. However, towards the end of the century the amount seemed to decline slightly, as it has been estimated that just over 5000 tons was the annual average weight exported from 1771 to 1778 inclusive.

The trade in the cured product was thus well developed and any fluctuations in the salt duties seriously affected the industry. Some fluctuations of the duty are briefly noted here:

1730. The salt laws were suspended.
1732. Salt duties were permanently reimposed.
1803 to 1815. This was the period of the Napoleonic Wars and the salt duty had been raised during this time until it reached a peak of 15*s*. a bushel.
1822. The duty was cut to 2*s*. a bushel.
1825. The salt duty was finally abolished.

The high duty on salt during the Napoleonic Wars had caused great distress throughout the West Country, as not only pilchard traders but also individual householders were affected by the duty. This seriously affected the poorer classes, because in order to survive the winter, a family needed to cure not only a considerable quantity of fish but also a pig. It was estimated that an average of 2250 tons of salt were used annually for the preparation of the salted product for the foreign markets, and about 350 tons of salt were used in Cornwall for salting fish for local consumption.

Throughout the historical records and, in fact, right up to the present, the pilchard has been notorious for its irregular appearances when considered over a period of years. For example, towards the end of the sixteenth century the main shoals appeared off the Cornish coasts in late summer and early autumn, but during the first quarter of the eighteenth century the shoals were not arriving until December. This affected not only the time when the main seining season occurred but also the quality of the fish differed when they were caught later in the year because of changes in fat content. By the third quarter of the eighteenth century they had again started to appear in their largest numbers in the summer and disappeared from the Cornish coasts by mid-October.

Ironically during the time of the French Revolution and the Napoleonic Wars (that is, from about 1789 to 1815) the fish were plentiful during most seasons, but because of the wars, the traditional foreign markets were closed and it was difficult to sell the fish. The situation was eased somewhat during this time by marketing certain by-products. For example, a demand came from the Navy for pilchard oil. About 10 gal of oil, which sold for £1, could be extracted from two hogsheads of pilchards. The fish were then sold off at 10*d*. per cartload for manure, and in August 1808 over 10,000 hogsheads of pilchards were sold, in Penzance alone, for this purpose. Before this time, however, pilchards had not been used so greatly as manure, as this was a more costly method per acre of ground than by the alternatives of sand or seaweed.

Because the wars had closed the European markets, attempts were made to find other outlets for the salted product. Some was exported to the West Indies, but these markets were not kept open for several reasons. First, it took some years for the tastes of the negro slaves to change to a preference for pilchards after North American cod and herring. Secondly, because of the longer voyage required to reach the West Indies, more salt was needed in the curing. Thirdly, for this particular market the fish had to be packed in 32 gal casks instead of the normal 50 or 52 gal sizes which were used for the Italian market. This meant that the curers had to change their methods; something that was alien to the nature of the Cornishmen as much then as now. The fourth reason why this market was not persevered with was that exports to the West Indies were shipped from Plymouth. This apparently worried the Cornish people as they felt that they might lose trade to the "foreign" Plymouth merchants. Finally, the chief exporting concern at the time did not like dealing with the slave-owning planters of the British West Indies who, moral considerations apart, were reputed to be slow in paying their debts.

The traditional method of catching pilchards was by seining, an ancient method of fishing thought to have been introduced to this country by the Carthaginians, as it is very similar to the method used in southern Spain and Portugal where the Carthaginians had bases. In the West Country the seine (or sean) net had been widely used for several hundred years until the early years of this century, but since the time of the First World War it has not been used at all for catching pilchards. A map illustrating the distribution of the seining stations of Cornwall during 1877 (Bennett, 1952) is shown in Fig. 6.1. It was not until about this time that the decline in seining began, but it soon became very rapid. The reasons for the decline will be dealt with later.

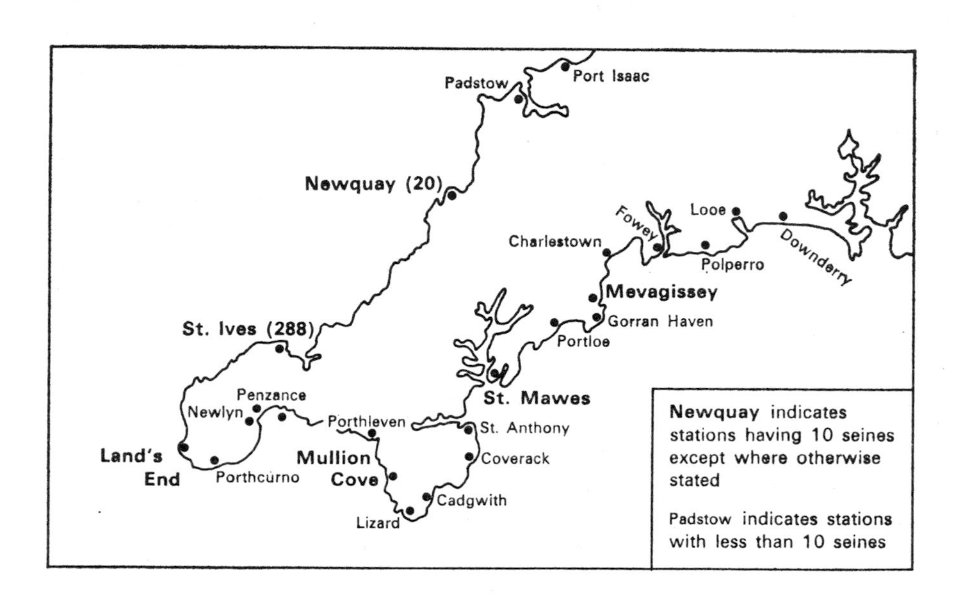

FIG. 6.1. Seining stations in Cornwall in 1877.

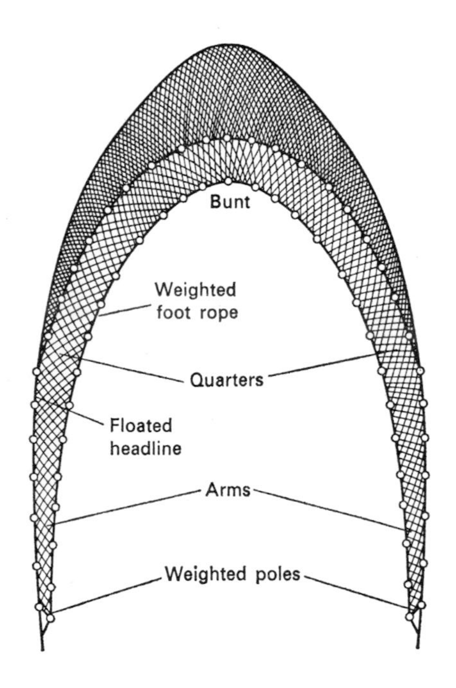

FIG. 6.2. Generalized shore seine to illustrate the Cornish pilchard seine.

The Cornish pilchard seine itself (Fig. 6.2) was of a minimum size of 180 fathoms in length and 8 fathoms in depth at the bunt and 6 fathoms in depth at the wings. This is according to the regulations set down at the chief seining station, St. Ives; it could apparently be as large as convenient. Like the drift nets of today, the seines were set in approximately by the third; that is, 18 ft of net were set onto 11 ft of backrope and 10 ft of footrope. The mesh size, introduced from the Dungarvan area of Ireland and known as the Dungarvan mesh, was small, having only 50–55 rows to the yard (Davis, 1937).

Catching pilchards by seining involved quite a number of men and boats, and the procedure was as follows. The net was shot from a large rowing boat, the net being paid out as the boat was rowed round in a circle. The circle was completed, if necessary, by a smaller net known as a stop net, 70 fathoms in length, placed in position by another boat. At St. Ives the team generally consisted of the large boat, two boats known as tow boats and a fourth, smaller boat. This fourth boat carried the master seiner and was known as the lurker. At most stations men were posted on the cliff to watch for fish and signal to the boats by means of a white ball on the end of a staff. These were known as huers, had exceptionally

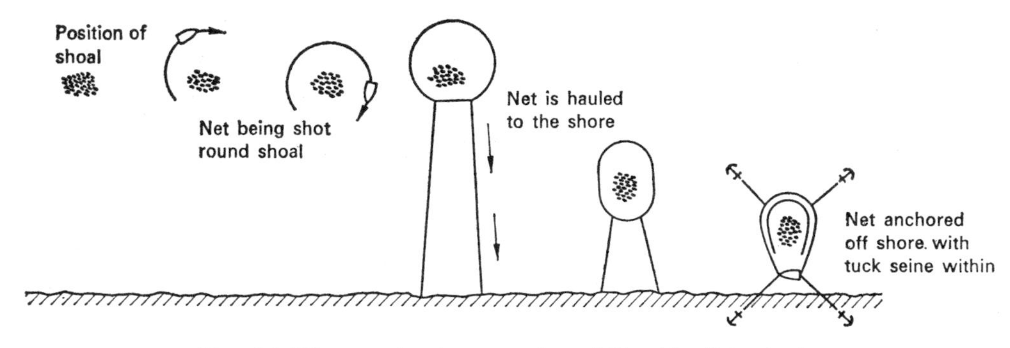

FIG. 6.3. Diagram showing operation of Cornish pilchard seine.

keen sight and, by their signals, directed those in the boats to the fish. They were allowed to hue from other people's property without being liable to action of trespass, by an Act of Parliament of James I's time. Probably because of the strain inherent in their work they were only on duty for 3 hours at a time. The number of men involved in the whole operation must have been about fifteen.

After the net had encircled a shoal it was then carefully towed into shallower water. Because of the necessity for this action it was sometimes known as a long-hauling seine. During hauling, boats not engaged in the process were rowed round the nets with their occupants splashing and throwing stones at the fish to keep them near the centre of the seine net. Large numbers of fish were often seined at a time so that it might require several days to remove them all from the net. The latter had, therefore, to be anchored in the water, and this was done by grapnel anchors or greeps. The fish were removed by long-handled nets or by shooting a smaller net, known as a tuck seine, within the circle. The tuck seine was hauled into an anchor boat rather than to the shore (Kennea, 1957). The operation of the Cornish pilchard seine is illustrated diagrammatically in Fig. 6.3.

Other men whose job entailed helping to heave the net in with capstans, also helped to carry the fish to the curing cellars. These men were known as blowsers. The method of pay for the men at the time of Wilcocks's report (1883) is interesting. The huers received £3 per month and one hogshead of every hundred taken during the season by the boats they worked with. The men on the boats received £2 5*s*. 0*d*. per month each and one-ninth of the fish between them.

It is interesting to note that, about the time that seining was at its height of popularity, in 1868 the greatest catch ever recorded being encircled by one seine was made at St. Ives: 5600 hogsheads were obtained which means (as a hogshead held about 2500 fish) that at least 14 million pilchards had been trapped by the net. The catch realized between £11,000 and £12,000 (Cornish, 1884).

The part played by St. Ives in the history of the pilchard industry seems puzzling. About the middle of the eighteenth century Mevagissey was the centre of activity of the industry. Exports from the area were 40 per cent of the total, only 4 per cent being sent from St. Ives. One hundred years later, in spite of the irregular appearances of pilchards along the west coast of Cornwall, St. Ives was becoming increasingly important in the industry as a seining centre. Merchants used the industry as an investment and bought whole seining outfits, or shares in them, and paid people to operate them. The industry became known as the rich man's fishery, a complete outfit cost about £2000 as compared with about £250 for the drifting equipment.

Of the 200 Cornish seines recorded in 1853, 132 were at St. Ives. However, apparently nearly half this number were unfit for use. Thirty-five years later, in 1870, the number of seines recorded had nearly doubled. There were then 379, of which 285 were at St. Ives and a maximum of 288 at St. Ives was reached in 1877. Mevagissey then had only ten recorded, a decrease of thirty in 35 years.

Another strange fact is that, in spite of seining being such an ancient method of fishing, it was not introduced to Looe until 1778, nor to Polperro until 1782. By this time, however, competition between seiners and drifters had already started, and the decline of seining was imminent.

As early as the beginning of the seventeenth century, the seiners had complained that drifters diverted and broke up the shoals, preventing them coming close enough to the shore to be taken by seine nets, and parliamentary action had been taken to legislate against the "growing evils caused by Driving Nets" and to prohibit the use of drift, trammel and stream nets within one and a half leagues of the coasts of Devon and Cornwall from 1 June to 30 November in each year.

(A complete account of the drift-net method of catching pilchards is given in Chapter 8.)

Despite this parliamentary action, drifters gradually gathered more adherents and from about 1750 to 1880 or 1890, the number of drifters and the number of men whose employ-

TABLE 4. FIGURES INDICATING THE DECLINE OF THE SEINE FISHERY AND THE INCREASING IMPORTANCE OF THE DRIFT FISHERY FROM 1785 TO 1870
(after Rowe, 1953)

Date	Number of drifters	Number of men employed	Number of seiners	Number of men employed
1785	180	900	110 to 140	1870 to 2380
1827	320[a]	1599	160[a]	2672
1835	360	1600	200	3400
1870	635	2462	379	1510

[a] Estimate.

ment was connected with them, increased steadily, even though more men were inevitably employed in a seining unit than in a drifting unit. Records are few, but Tables 4 and 5 give some indication of the decline of the seine fishery and the growth of the drifter fishery in Cornwall.

Table 4 shows the large increase in the number of men employed in drift fishing over the 85-year period. It also shows the large increase in the number of drifting units that occurred during the period. The number of seines is also seen to have increased at the same time but many of these would not be in working order. The owners, by investing in more than one seining outfit, were entitled to more fishing rights, even though they might only be operating one, provided that the outfits were still registered. This explains why the number of registered seines increased whilst the number of operational units decreased. The distribution of the seines and drifters at various ports in 1870 is shown in Table 5.

TABLE 5. DISTRIBUTION OF THE FISHING OUTFITS AT THE MAJOR PORTS, 1870 (after Rowe, 1953)

Number of seines	Based at	Number of drifters	Based at
285	St. Ives	176	St. Ives
10	Mevagissey	130	Newlyn
9	Newquay	105	Mousehole
9	St. Mawes	104	Porthleven
43	Other ports	61	Mevagissey
23	Out of commission	22	Looe
		37	Other ports
379 total		635 total	

The last quarter of the nineteenth century saw the catches from drifters exceed those from seining, and from this time the decline in seining became more marked and rapid, Eventually, by 1920, as catches using seine nets had not recovered since the First World War, this type of net dropped from use completely.

This decline can be attributed to several causes. Perhaps the primary cause was the infrequence with which the fish appeared in the bays. This led to a great deal of uncertainty in catching the fish by seining and favoured drifting. An historical fact, which was rather an anomaly, and which favoured drifting rather than seining, was that catches in drift nets were exempt from tithes after 1830 but tithe duty was payable on fish caught by seining. Fishermen in some regions, for example Paul and Mousehole, had refused to pay tithes before this on fish caught by drifting even though their refusal resulted in involving them in considerable local struggles. Among other reasons for the immunity to tithe payments of drift-caught fish was the fact that the distance from the shore at which the drifters operated was generally beyond parochial jurisdiction. Furthermore, when the fish were pulled from the water in the drift nets they were dead and this also meant that they were exempted from tithe duty.

Another important factor which affected the basic organization of the fishing industry was the tendency, towards the end of the nineteenth century, for larger ships to be built. Dr. A. J. Southward has suggested (personal communication) that these larger boats were built not for pilchard fishing but for trawling. This enabled the fishermen to obtain a steadier living as they were then able to catch fish which commanded a better price than did pilchards. However, much of their trade would rely on the establishment of rapid channels of distribution for the fresh fish and this would not have been possible without the advent of the railway. The more or less simultaneous development of trawling and rail transport in the area had its effect on the pilchard industry not only in the redirection of fishing effort into trawling but also in diminishing the local importance of many of the minor seining stations situated around the coast. This was, at least partly, because the larger vessels needed larger ports and only a few of the seining ports were capable of extending. Furthermore, it was economically possible to have only a few large ports where formerly there had been several small ones. The ports which developed to accommodate the needs of the larger vessels gradually became more important.

It is interesting that the numbers of both seines and drifters increased quite noticeably during this time and it is possible that this may have been partly due to the opening of the railway. This could have been advantageous to them in that the markets for pilchards suddenly became considerably larger instead of very localized as formerly. However, this phase did not last long for the seiners which soon went into an irreversible decline. It is not known for certain whether the rail influenced this, but the fact that the main seining station, St. Ives, was not connected to the railhead may have been a contributing factor.

Finally, the numbers of boats and men in a seining unit were very much greater than for a drifting unit, which although it needs up to twenty-five nets, employs only four or five men on one boat. Shore seining was essentially a part-time occupation which might pay well for a short period, but which could no longer maintain itself with competition from the virtually full-time drifters. Also, as mentioned earlier, the initial cost of the drifting outfit was about one-eighth that of the seining outfit. These factors would influence men embarking on a fishing career to take up drifting rather than seining.

It can now be seen that the changes which have taken place in the English pilchard industry cannot be said to constitute an advance in the structure of the industry. The change from seining to drifting has been due to economic and biological factors and the technical advances on the processing side have been accepted by the industry because of economic necessity. However, these changes cannot be considered to be more profound than adaptations to changes in the environment. That is, in spite of these developments the pilchard industry has shown no signs of growth (if anything, there has been a decline of its influence) and it has remained on the same level of industrial organization for at least 250 years. This state of retarded growth has been due, at least in part, to the fact that it has been guided by no conscious policy and has, therefore, remained at this primitive stage of organization.

CHAPTER 7

STATISTICAL SURVEY OF PILCHARD CATCHES IN RECENT YEARS

The Period through 1965, with a Brief Survey back to 1885

Figure 7.1 illustrates the landings of pilchards in the English Channel for the post-war period. The statistics (Great Britain, Ministry of Agriculture, Fisheries and Food, 1945–66) give the total quantity of pilchards landed in Great Britain as well as those caught in the Channel and landed at the ports in Cornwall. Only the quantities landed at the Channel ports are shown in Fig. 7.1 as the pattern for both groups of landings is determined by the

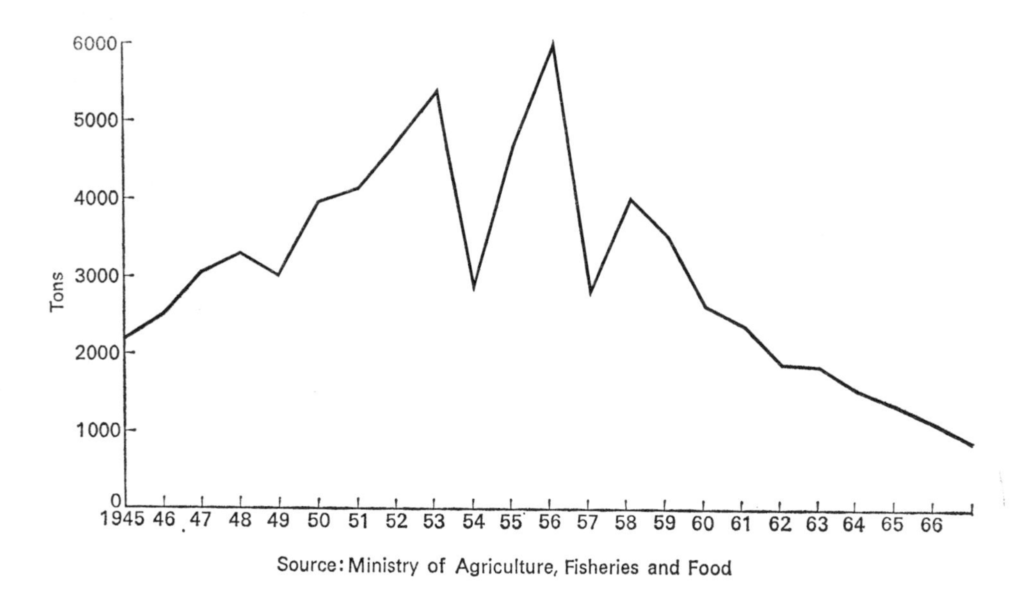

FIG. 7.1. Pilchard landings in post-war period from the English Channel.

Channel fishery and the only year in which the two quantities differed markedly was 1951. In that year a considerable quantity, relatively speaking (1670 tons), was caught in the North Sea. According to the statistics not more than 200 tons of pilchards per year are now caught by vessels in the North Sea, although more than this may be taken incidentally with herring. Thus, rather more pilchards may come from this area than are actually recorded but this quantity would not make any significant change to the pattern given by the recorded statistics. Table 6 gives the data for the whole of Great Britain.

TABLE 6. LANDING OF PILCHARDS IN GREAT BRITAIN FROM 1945 THROUGH 1966

Year	Quantity Tons	Year	Quantity Tons
1945	3200	1956	5946
1946	2626	1957	2957
1947	3154	1958	4148
1948	3450	1959	3617
1949	3020	1960	2986
1950	3935	1961	2700
1951	5805	1962	2018
1952	4053	1963	1966
1953	5456	1964	1724
1954	3001	1965	1348
1955	4850	1966	1132

Considering the catches made in the Channel as being typical of the major trends throughout the country it can be seen from Fig. 7.1 that there is no consistent sequence of events before 1958. From 1946 through 1952 there was an overall tendency for the catches to increase in size, there being a slight decrease in 1949. The first 3 years of the post-war period through 1948 showed a gradual improvement with the 1948 catch reaching 3352 tons. Greater catches were not really expected at this time but the fall in 1949 was disappointing in an industry where steady improvement after wartime conditions should have occurred. However, in 1953 over 5000 tons were caught and it appeared that the industry was returning at least to pre-war conditions.

Unfortunately, 1953 proved to be the last year of the upward trend in the landings and, except for 1956, had the highest catch in the 21 years being considered. Since 1953 there has been a mean downward trend in total catch and the annual catches have undergone considerable fluctuations. For example, in 1954 the catch fell to 2986 tons, a figure below even the immediate post-war totals; but in 1955 and 1956 there was marked improvement, the Channel vessels landing 4720 tons and 5890 tons respectively. Unpredictably, this good year was followed by a very great fall in catch, as the quantity landed was even lower in 1957 than the 1954 total, amounting to only 2844 tons. In 1958 an improvement was made when about 4020 tons were caught. Nevertheless, this improvement was only temporary because since 1958 the catches have shown a steady decrease which can be seen in the graph and Table 6. (1967 landings were 711 tons.)

Apart from the recent downward trend, however, there has been no other consistent pattern of landings of British pilchards. An attempt may, therefore, be made to explain some of these wide fluctuations. First, it should be mentioned that it is not the total biological resource that is so variable, there being an estimated 800,000 tons basic stock of pilchards in the Channel (Cushing, 1957), but as long as the industry remains at its present, primitive stage of development (as discussed in the previous chapter), there are likely to be substantial yearly variations in the proportion of this stock available to them.

The amount caught per year must, therefore, be controlled largely by factors other than biological. An example of the non-biological influences can be found if the imports of pilchards are studied in connection with the British catch. Up to and including 1954, the imports of canned pilchards and the British catch varied inversely (Fig. 7.2). However, since 1954 the two elements have moved together; the catch and the imports rising or falling at the same time. This suggests that the market was saturated by 1954 and that there was a more or less tacit agreement by the buyers of the big chain stores to order local and South African pilchards in a given ratio. The rapid development of self-service stores had led to increased consumption of canned goods (see Chapter 11) and with proper advertising the total consumption of canned pilchards may again start to rise. The imports are mostly from South West Africa, although small quantities also come from Japan and the Union of South Africa.

1885–1945

For the purpose of illustrating the catches graphically (Figs. 7.3 and 7.4), the English landings of pilchards from Cornwall for this time have been divided into those from the south coast and those from the west coast of the peninsula. However, since 1908 catches

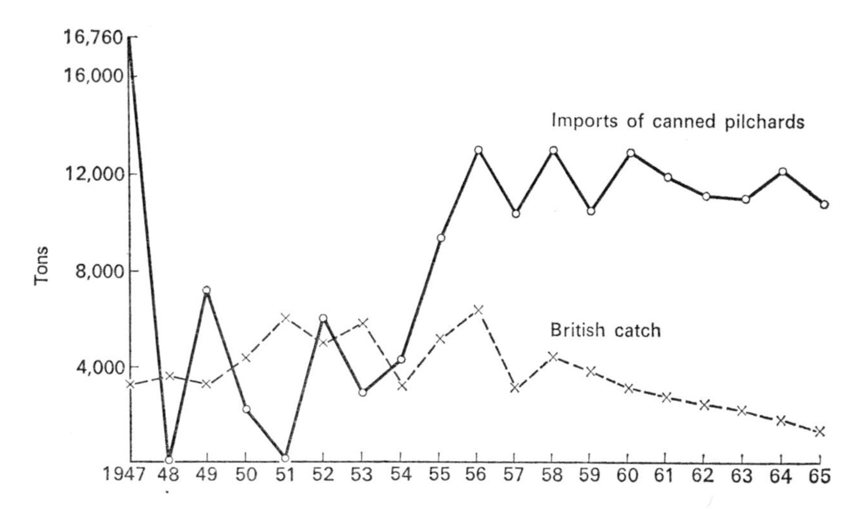

FIG. 7.2. Imports of canned pilchards compared with British catch.

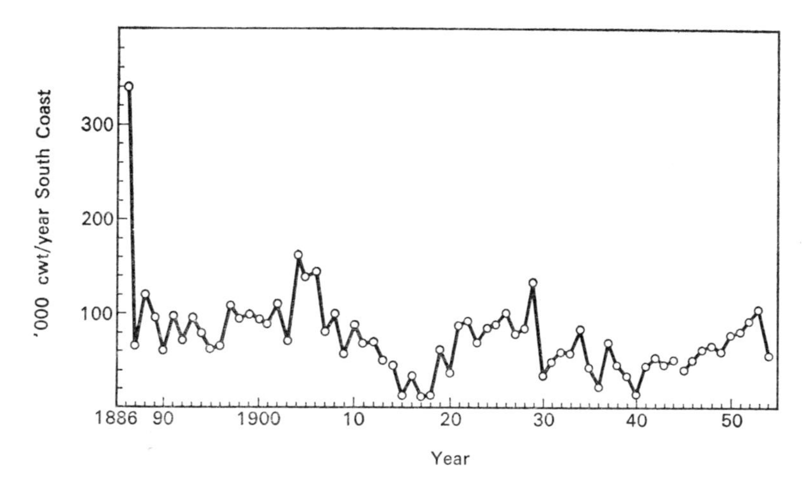

FIG. 7.3. Landings of pilchards on south coast of Cornish peninsular, 1886 to 1954.

from the west coast (which really means from Sennen Cove and St. Ives) have been very small indeed. This is because the shore seine which was in operation at these ports has not been used for pilchards since the First World War period, and for some years before this the catches were tending to be small.

On the south coast, from 1885 to 1912 the catches never fell below 3000 tons. There was a period of exceptionally high catches from 1904 to 1906 in which the catch averaged about 6350 tons. Although this quantity is greater than any reached since the war, it must also be remembered that the total quantity taken from the Channel was even greater, as the west coast landing for 1905, for example, totalled 1520 tons. With the catch from the south coast of 1905 being about 6090 tons, landings from the West Country for that year reached over 7600 tons. In general, however, the quantity from the south coast averaged about 5000 tons.

The graph shown in Fig. 7.3, giving the catches for the south coast, shows that catches fell, as would be expected, during the period 1914–1918 and again during the period of

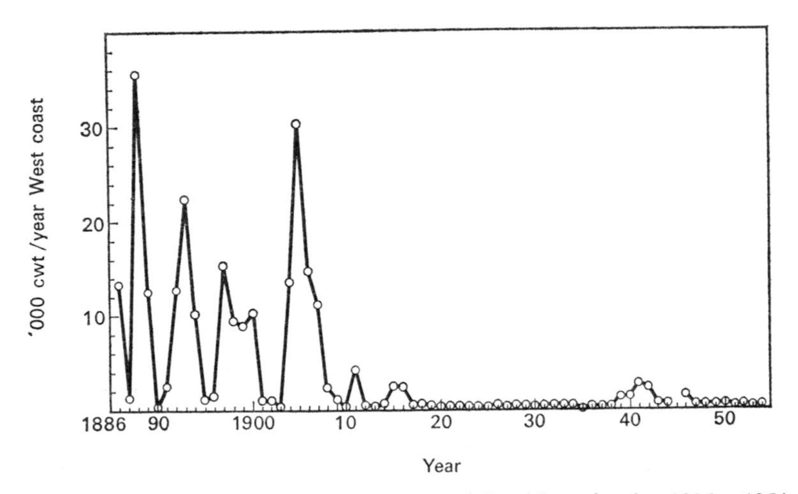

FIG. 7.4. Landings of pilchards on west coast of Cornish peninsular, 1886 to 1954.

economic crisis around 1930. In fact from 1930 until after the Second World War there were only 2 years during which Cornish catches of pilchards rose above 3000 tons. These 2 years were 1934 and 1937, when the catches totalled about 4720 tons and 3300 tons respectively.

The graph in Fig. 7.4 shows the west-coast landings. As already stated the catches from this area dropped to an insignificant quantity after 1916 because of the disappearance of the shore seine. For the years before this date the catch fluctuated violently. The total averaged about 660 tons, but this was the total for only two ports (i.e. Sennen Cove and St. Ives) and compares favourably with pre-war landings at Looe and Mevagissey. At recurrent periods of every 3 or 4 years the catches became high, reaching above 1500 tons. It is interesting to compare this size of catch, from just two ports, with the post-war average from the whole area which is equal to little more than twice this figure, namely 3120 tons.

It is evident that the fall in total catches of later years compared with pre-1910 has been due, not only to changes in technical matters, but to a marked fall in demand for both fresh and salted pilchard. Previous to the First World War the latter was a main food staple in the West Country of England, and there was considerable local demand. Also the product was eaten quite widely in Italy, the two markets creating a total demand which probably could only be met by the Cornish industry. Now the demand for pilchards in Britain may be as great as before, but it is the canned product which is demanded and this is largely met by imports. That is, the onus for fulfilling the demand no longer really rests with the Cornish pilchard fishermen.

Analysis of Monthly Catches from 1947 to 1965 Inclusive

The monthly catches for these years have been illustrated in block diagram form and are given at the end of this chapter.

Two main trends emerge from a study of these figures. These are:

1. a change from former large November and December catches to rather poor catches in these months;
2. a general levelling out of the catch over a longer period of the year from June to the following January.

These two facts will now be dealt with in more detail.

From 1947 through 1952, November could always be relied upon to produce the largest total of any month. However, in 1953 the November catch was less, but only marginally so, than the December catch for that year, and in 1951 the November catch was exceptionally large, being about 3270 tons. Since that year it has never risen above about 560 tons. However, it was during that year that the unusually large landings from the North Sea occurred and these, no doubt, contributed to the size of the catch for November 1951. Before 1953 the January figures, with the exception of 1950, were below about 250 tons, but since then they have frequently risen to 500 tons or above. A change occurred after 1953, which altered the catching-pattern so that November catches became lower than they had been previously. From 1953 through 1960, with the exception of 1959, December catches tended to improve on the November figure and this trend frequently continued into the following January. For the last 4 years, from 1962 through 1965, catches during November and December and during the first 3 months of the year have fallen rapidly to become negligible. The block diagrams of the monthly catches for 1963, 1964 and 1965 show another pattern of landings beginning to emerge.

The correlation of this change in catching pattern in 1953 with the beginning of the downward trend in landings is interesting. The industry relied on the catches in the late months of the year to bring the landings up to what was its average level. Now that fish can no longer be regularly caught in the last quarter of the year the industry has gradually decreased. Nor is it possible to catch the fish during the first 3 months of the year as they are not available in numbers which make fishing a profitable proposition.

The change seems to be due to a change in the habits of the fish which no longer appear near the shore, in the numbers of former times, during the last quarter of the year.

With the exception of 1952 the figures for catches during the months of June, July, August and September were generally low until 1954. The years 1953 and 1954, however, did show slight improvements of size of catch during August and September. From 1954 through 1965 the summer months showed much better catches and these tended to be of a similar size for each month, although 1960 and 1962 did show a gradual increase throughout these 4 months, and 1957 figures were low. From 1958 through 1962 the catches for August and September were amongst the best for the whole year; and in 1961 the August total was the largest, reaching 636 tons.

Further facts also emerge as corollaries of these two main trends. One of these is that the catches for February, March, April and May tend to be low, although there are exceptions to this; for example, in 1953 the catch for March was unusually high, reaching about 490 tons. Also in 1955 and 1956 the April catches were high, being about 570 tons and 730 tons respectively. Catches are low at this season of the year because, as stated earlier, the fish are not usually present in the Western Channel in commercial quantities.

Strange anomalies in fluctuations of catch size are also brought out by a study of these monthly figures. For example, in October 1960 the total catch was just over 3 tons, although

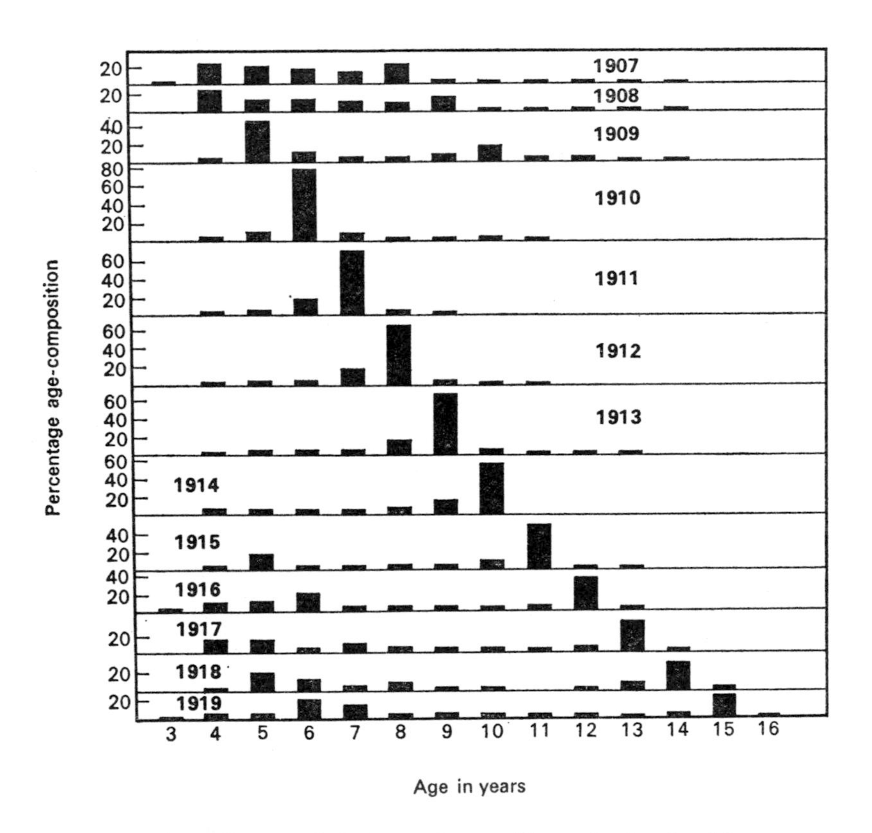

FIG. 7.5. Influence of 1904 year-class on Norwegian herring fishery.

the average for the 5 years previous to this had been about 280 tons. Again one would normally expect the catch for July to be in the region of 500 tons. However, in July 1957 only 138 tons were caught. This tendency to severe fluctuations in catch size for any given month makes the chance of giving accurate forecasts from month to month extremely low. Such catch forecasts would be of use to the industry in programming it from the catching side through to marketing. It would be of especial use to the canners, who would be able to plan their intake and, for example, make suitable arrangements to obtain alternative raw material or cut down staff hours if the forecast were low.

Forecasting

It occasionally becomes apparent that established fisheries fail for reasons which, although obscure, are probably a result of the combined effect of natural and economic causes working to the detriment of the stock. In the past these failures have occurred most frequently in pelagic fisheries. For example, in 1898 the French sardine fishery yielded 100,000 tons, but in 1902 there was a yield of only 18,000 tons. In 1920 the British mackerel fishery in the North Sea failed and, more recently, the U.S. Pacific sardine fishery failed, falling from 320,000 tons in 1950 to 6500 tons in 1952 (Simpson, 1956).

There are also smaller fluctuations which can be caused by a variety of factors such as variation in effort and introduction of different types of gear. Nevertheless, even if effort

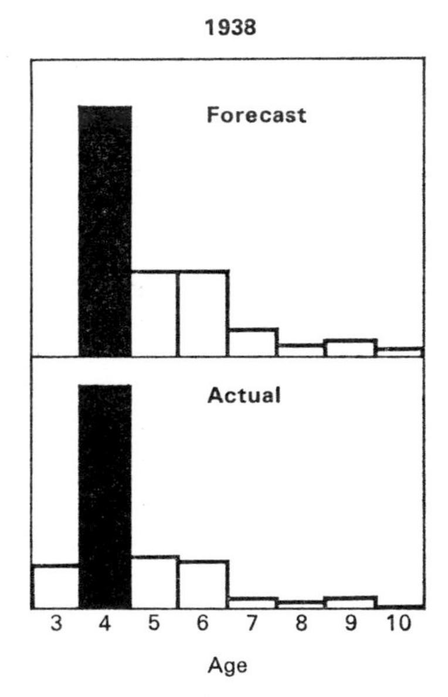

FIG. 7.6. Herring fishery of East Anglia: forecast and result, 1938.

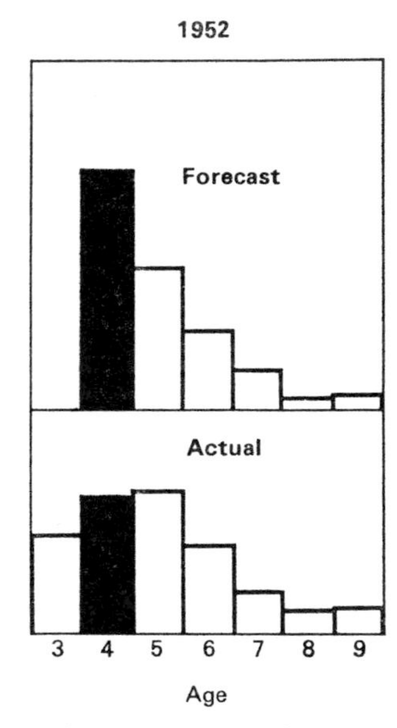

FIG. 7.7. Herring fishery of East Anglia: forecast and result, 1952.

and gear used remain more or less constant, considerable fluctuations remain in certain fisheries, and the pilchard is the least predictable of all important fish. Such fluctuations could be caused by external factors which, by their variation, alter the ease with which the fish are located and caught. Such factors are winds, currents, visibility and so on. However, working with the Norwegian herring, Lea and Hjort (quoted in Simpson, 1956) found that fish derived from the spawning of one or two particular years may influence the catch over a large number of years. In the Norwegian herring fishery the individuals surviving from the spawning of 1904 dominated the fishery for a number of years. The influence that these fish had on the stock for 12 years is shown in Fig. 7.5. They came into the fishery as 4-year-olds in 1907 and from then on dominated the stock until 1919. The fish spawned in 1904 were known as the 1904 year-class and this introduced a new idea into the study of fish populations.

This was the concept of year-classes and it has been extremely useful in forecasting in certain fisheries. A long-standing and detailed knowledge of the stock is, however, necessary before forecasting from year-classes can be successfully achieved. Even then all the research that has gone into making forecasting possible can be rendered useless if, for example, a foreign fishing nation changes its methods of catching the species in question. This happened in 1951 in the herring fishery of East Anglia. In this particular instance Hodgson (1957), analysing the composition of the catch with respect to year-classes, had been providing forecasts, some very accurate indeed, for about 30 years (see Fig. 7.6 for 1938 forecast and actual result). He had also investigated the effects that the moon had on the size of the catch through its influence on fish behaviour. He found some interesting correlations and with them was able to increase the accuracy of his estimates, and also provide an indication of the time of the onset of the fishery in East Anglia.

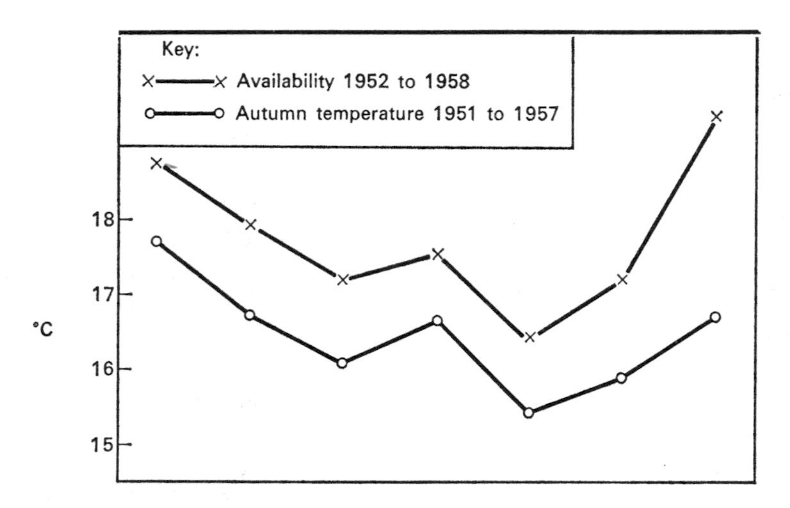

FIG. 7.8. Correlation between autumn surface temperature and pilchard availability for South African pilchard.

In 1952, however, the forecasts broke down so completely that their publication was suspended (Fig. 7.7). The failure of this forecasting method was formerly ascribed to the rapid increase in fishing effort that had occurred in the North Sea at about this time due to industrial fishing for young herring. It is now thought that complex biological changes were the cause of the failure of the East Anglian herring fishery.

A method of forecasting used with considerable accuracy for the South African pilchard, and not requiring any empirical knowledge of the stock, was applied to the Cornish pilchard. It was in 1957 that Buys (1959), working with the South African pilchard, *Sardinops ocellata*, found a remarkable correlation between the autumn temperature of the sea in the 0–50 m layer, in the areas of the fishery of the Atlantic Ocean off the coast of South Africa, and the size of the catch in the following year (Fig. 7.8). Taking autumn temperatures as those between August and November inclusive, the principle has been applied to certain areas of the Western Channel and the pilchard landings of England. The temperatures of these 4 months were averaged for each year from 1919 to 1927 and the catches for the years 1920–28 were plotted on the same graph. However, it was found that even out of these 9 years two of them exhibited an inverse correlation, so the method was not deemed satisfactory for the English pilchard fishery.

It seems, therefore, that at present there is no method of forecasting which can be applied to the English pilchard fishery. If more were to be known of the age composition of the stock, then forecasting would become a possibility.

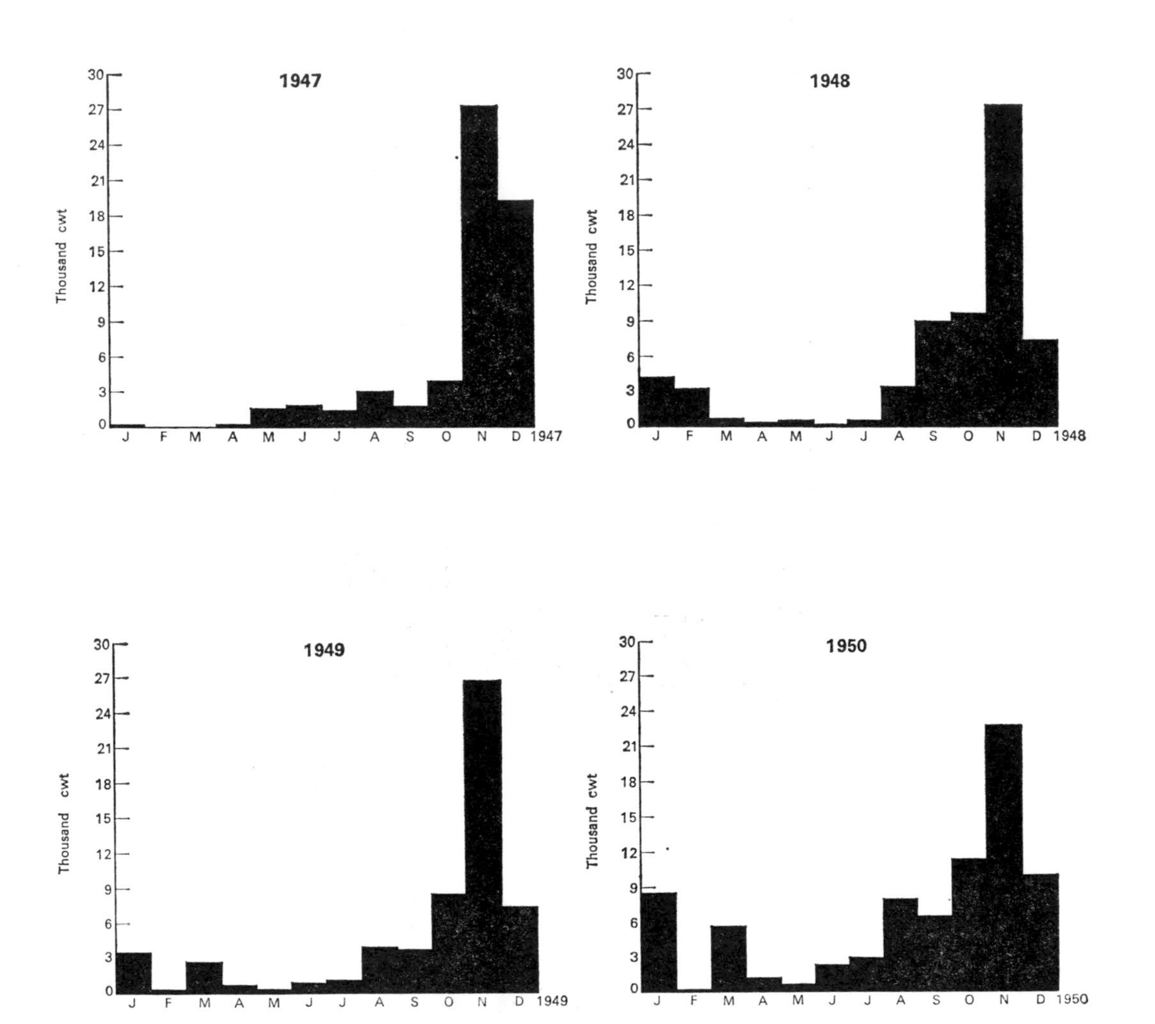

FIG. 7.9. Histograms of monthly catches of English pilchard, 1947 through 1965.

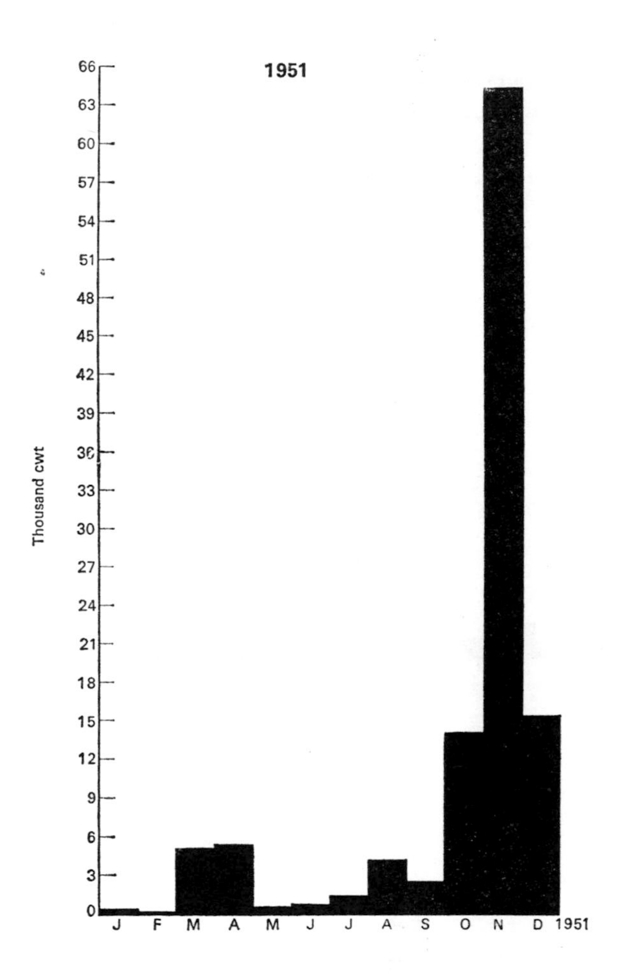
1951
Thousand cwt
66
63
60
57
54
51
48
45
42
39
36
33
30
27
24
21
18
15
12
9
6
3
0
J F M A M J J A S O N D 1951

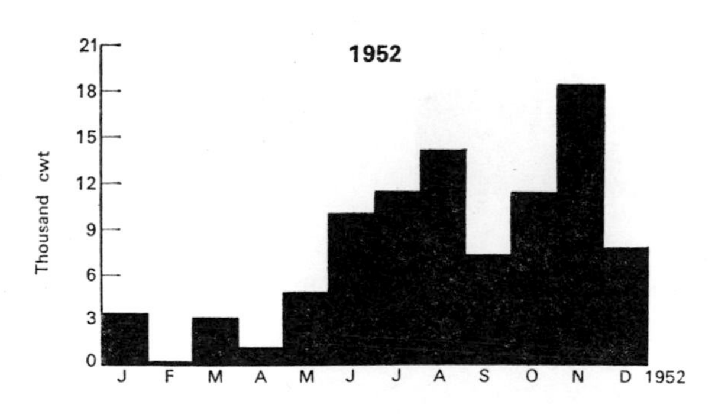
1952
Thousand cwt
21
18
15
12
9
6
3
0
J F M A M J J A S O N D 1952

FIG. 7.9.

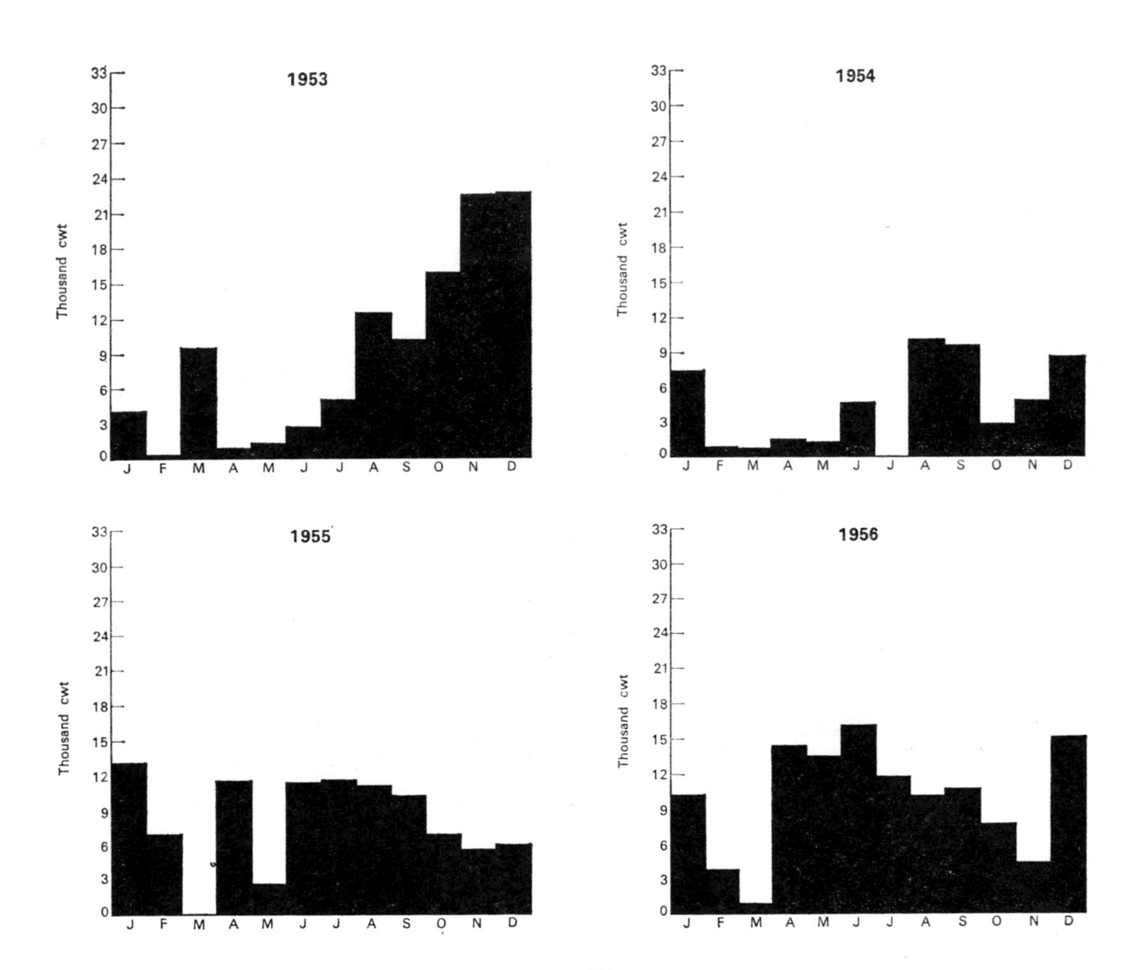
1953
1954
1955
1956
Thousand cwt
33
30
27
24
21
18
15
12
9
6
3
0
J F M A M J J A S O N D

FIG. 7.9.

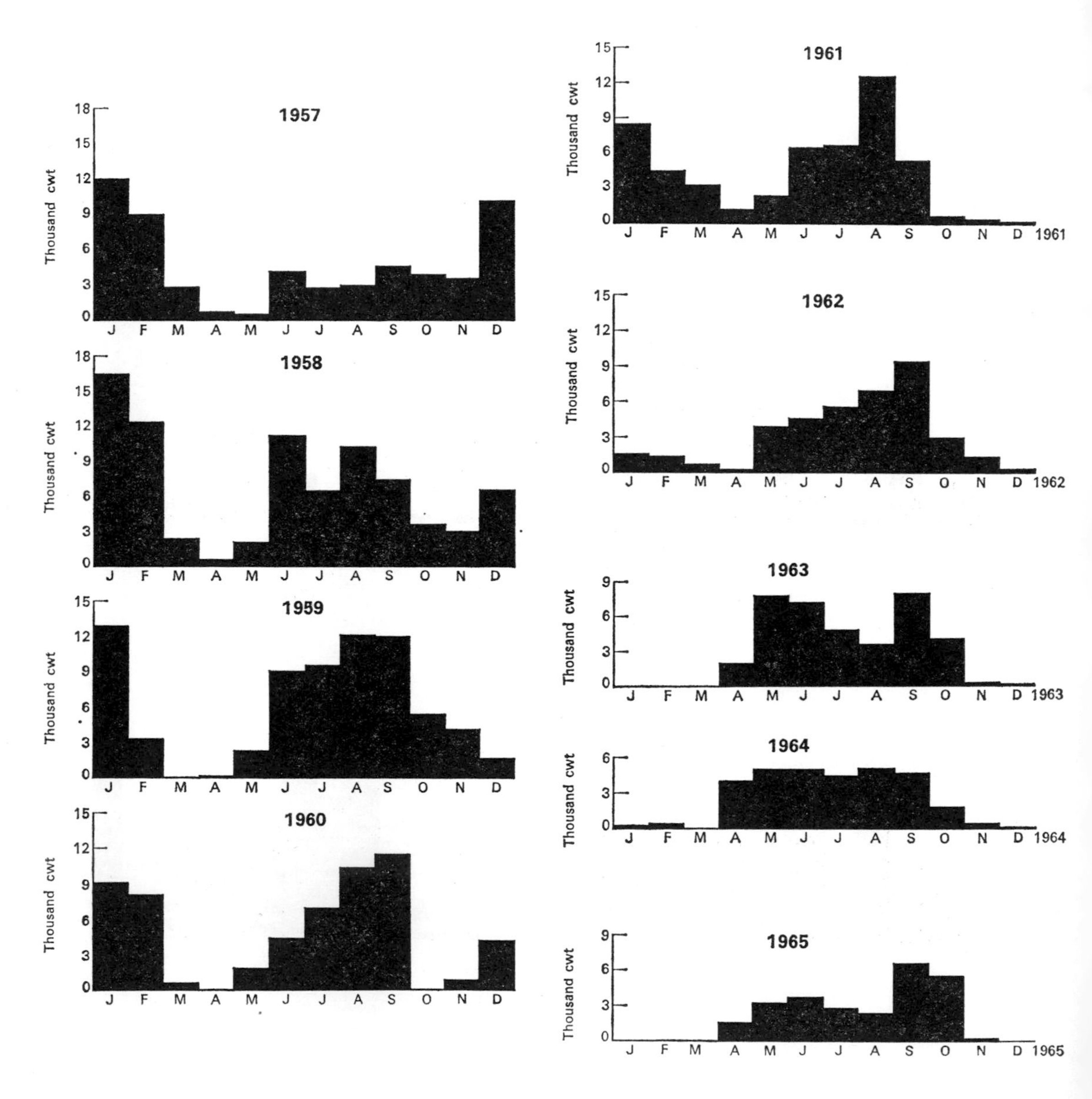
1957
1958
1959
1960
1961
1962
1963
1964
1965
Thousand cwt
J F M A M J J A S O N D

FIG. 7.9.

CHAPTER 8

CATCHING TECHNIQUES

AT THE present time only one method of catching fish is used by the pilchard fishery of Cornwall. This method involves the use of a considerable length of net (up to $1\frac{3}{4}$ miles) and a vessel usually 35–45 ft long. The vessel is known as a drifter and the net as a drift-net. The fishing operation is usually known as pilchard drifting but sometimes as pilchard driving. The fishing vessel leaves the home port some time before sunset (not usually more than three-quarters of an hour) and spends up to $1\frac{1}{2}$ hours reaching the area and searching for fish. Location of the fish is still largely carried out in the first instance by the use of "natural signs" such as a large flock of gulls due to a mass of fish in the area, or the surface of the sea broken by bubbles which make the water paler in colour and also indicate the presence of fish. Echo-sounders are used also but, although the skippers seem to enjoy using them, the machines are often run in a poor state and fishing will sometimes be started without any definite trace from a shoal or traces from individual fish being recorded on the sounder. Also if one member of the fleet reports over the radio that he has a trace on his echo-sounder, several other members of the fleet are likely to move into his area and start fishing without seriously checking the report. The "feeler wire", which was once used and was quite valuable for checking for a shoal, does not appear to be used any more .The "feeler wire" was a piece of piano wire about 6 ft long with a weight at one end. This was dragged over the side of the boat and skilled operators could feel the fish bumping against it. From the number of bumps the wire received it was possible to tell if the shoal was dense enough to make the shooting of the nets worth while.

Once the skipper has decided where to fish, the operation known as "shooting" of the nets takes place. The sail of the drifter is hoisted and she starts to sail downwind. If necessary the engine is used to give her enough way to clear the nets. The first net is thrown overboard, and attached to it is the second net to which is attached the marker buoy. This buoy is lighted and is known as a "dhan". After this the nets, fastened to one another, are thrown over the starboard side in rapid succession. A full complement of nets is twenty-five. This is known as a full fleet, but at the extremes of the season it is not often that more than eighteen nets (and frequently even less) are used. At the maximum this wall of net is $1\frac{3}{4}$ miles in length, but this may vary from port to port. When the final net is shot the warp, or messenger rope attached to the top of the nets, is made fast to the forward winch and enough is paid out to allow the ship to ride easily at the nets. Her head then comes round into the wind under the influence of the mizzen sail.

The boat then drifts, usually downwind, but if there is no wind she may drift with the current, for a time which varies from about 1 to 4 hours. It is interesting to note here that whilst they are in shoals pilchards have a tendency, which is in direct contrast to that of the herring, to move with the wind. That is, the pilchard is a "lee" fish whereas the herring is a "windward" fish. Hauling of the nets then occurs. The boat, under her own power, very slowly retraces her way along the path of the nets. This enables two men to haul in the nets by hand over the starboard side. The fish are shaken out into the hold and the nets restored in their place next to the hold. The time taken in hauling depends on the size of the catch, the number of nets, and the state of the sea. If the whole fleet of nets is out, then even if the catch is poor and the sea calm, the process is likely to take 4 hours at least. If the catch is good then hauling will continue until the early hours of the morning, the return to port being very early in the morning. In cases of very heavy catches the nets may be hauled and stored with the fish still gilled. The fish would then be shaken out on return to the port.

Nets

Drift nets of today are generally still made of cotton (although experiments with synthetic nets have been carried out which will be mentioned later) and are of a type which first came into use in connection with the herring industry around 1856. The standard herring

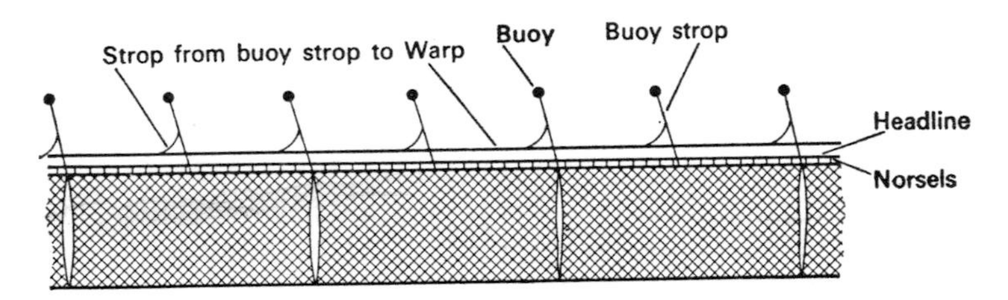

FIG. 8.1. Pilchard net.

drift net measures 33 yards when mounted on ropes, but 55 yards before mounting. It is 15 yards in depth. The slackness obtained on mounting increases the net's efficiency and pliability.

The netting is bounded by stronger material, the guard, which is finished off with cotton line. This in turn is attached by short lengths of twine called ossels to the cork line or top of the net, and to the sole rope or bottom of the net. The nets are attached to each other by means of a loop at each corner.

The pilchard net is basically the same construction as this. (Fig. 8.1.) There is a warp running above the headline which is sometimes called the messenger rope. Running from the headline at 5-fathom intervals are strops, on the end of each of which is a buoy. These strops are known as the buoy strops and they are 2–3 fathoms long. The reasons for their use are:

1. The nets are likely to catch more fish when the headrope is below the surface.
2. The nets are less likely to be damaged by the keels of other vessels when several feet below the surface.

The length of strop governs the depth of the net in the water and experience gives the best length of strop, although some experimentation may lead to a more satisfactory length being adopted under certain conditions. The other strops connect these buoy strops to the warp. The warp is thus supported at about the level of the headline.

The headline itself is composed of two ropes with opposite twists to prevent kinking. It is often corked, that is, has cork floats fixed between the two ropes, or on them, and it is attached to the net by means of short lengths of double-twisted twine. As mentioned above, these are called ossels on the herring net but nossels or norsels by the Cornishmen.

Pilchard nets are known as fly nets because of the lack of strengthening at the foot or side (selvage) of the net. Also their method of lying in the water enables them to be described as sunk nets as they are fished with the net below the warp. The herring drift net, on the other hand, has the warp or messenger rope, below the nets and is described as a swum net.

Accessory Gear

In order that shooting and hauling can be done efficiently and the nets positioned most advantageously in the water, certain accessory items of gear are needed. Most of these have been mentioned already, but they will be listed and briefly described here. They are: (1) stoppers; (2) buoy ropes; (3) messenger rope or warp; (4) canvas buoys or pallets.

1. *Stoppers* are used for attaching the nets to one another at the bottom. They are about 3 fathoms in length. (They are not shown in the diagram of the pilchard net.)
2. *Buoy ropes or strops* are the ropes extending from the nets to the buoys. They are 2 or 3 fathoms in length and if their length be altered the depth of the nets in the water can be changed.
3. *Messenger rope*. In the pilchard nets this is situated at the top of the net and is generally known as the warp. It is made of manilla and is from 3 to $3\frac{1}{2}$ in. in circumference. It is tarred and sinks below the surface, but is buoyed up by strops extending from the buoy strops to it.
4. *Canvas buoys or pallets* are used at intervals along the net to buoy it up. They, incidentally, help to mark the position of nets in the water. They are also known as buffs or bowls. The relationships of these parts to one another can be seen in the diagram (Fig. 8.2), of the herring net, included for comparative purposes.

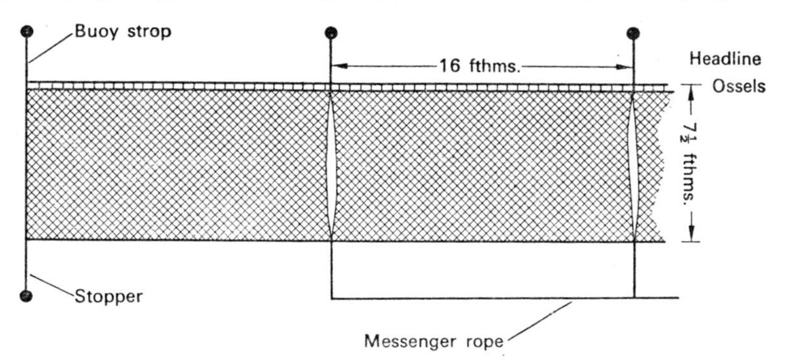

FIG. 8.2. Herring net.

Pilchard nets are always "set in by the third", that is the length of the rope on to which they are mounted is two-thirds the length of the net, so that the final overall length of the netting is one-third shorter than the actual net. Different lengths of net are used in different parts of the West Country. For example, at Newlyn a net is 87 yd long, mounted on 58 yd of rope; at Mevagissey the 120-yd-long nets are set on to 80 yd; at St. Ives nets of 100 yd were set in to rather more than two-thirds of rope—96 yd—and at Looe 180 yd of net are set on to 120 yd of rope.

This leads to the final length of the full fleet of nets varying as mentioned earlier. It can be seen that the final length is determined by the home port of the vessel. A full fleet is twenty-five nets. Thus at Newlyn the full fleet would be 1450 yd long or 0·8 of a mile. At Mevagissey, where each net is set on to 80 yd of rope, the length would be 2000 yd or just over $1\frac{1}{8}$ miles long. At St. Ives the final length would be 2400 yd or nearly $1\frac{2}{5}$ miles, and at Looe, with nets set on to 120 yd, the fleet would stretch for 3000 yd, or nearly $1\frac{3}{4}$ miles. This is summarized in Table 7.

TABLE 7. DIFFERENT LENGTHS OF PILCHARD DRIFT NET AT FOUR CORNISH PORTS

(Measurements in yards)

Port	Length single net/ length when set on rope	Number of nets in full fleet	Length full fleet
Newlyn	85/58	25	1450
Mevagissey	120/80	25	2000
St. Ives	100/96	25	2400
Looe	180/120	25	3000

The depth of the net is described by the numbers of rows of mesh between the headline and the bottom of the net. The number is usually given in scores. If there is any strengthening or "heading" to the net (where the meshes are often made double) then the number of rows of heading is given separately. At nearly all ports the depth is 16 score plus 10 meshes of heading, although at Newlyn the nets are 18 score deep plus a heading of 10 meshes. This works out at a depth of nearly $4\frac{1}{2}$ fathoms, or 5 fathoms at Newlyn, plus 10 meshes in both cases.

Mesh

There are several ways of referring to the size of meshes (Davis, 1958).

1. The one most frequently used now is the number of rows to the yard. The net is held out along a yard measure and the number of meshes to the yard counted. The nets used in the pilchard industry usually have 36 meshes to the yard when new (although after long use they may shrink to 37–39 rows to the yard). Davis, however, says that at Mevagissey 39 rows per yard is a standard measurement and at St. Ives 34–36 rows per yard is the standard.
2. This mesh of 36 rows to the yard means that each side of it is 1 in. long. This is sometimes therefore referred to as a 1 in. mesh or 1 in. "bar".
3. If the net were pulled so that the sides of the mesh were pulled together, then the mesh could be called a 2 in. mesh usually referred to as "stretched".
4. The distance right round the mesh, the sum of the four sides, can be used, although this is not a common method. Using this method the 36 rows to the yard would be called a 4 in. mesh.
5. Finally, the mesh size may be measured using some specialized local method, such as a ring gauge, when the size will then be referred to in some other way.

These different methods of measurement can lead to a lamentable state of confusion, as a measurement just stated as a 4 in. mesh could mean 36 rows to the yard using the fourth method of measurement, or 18 rows to the yard using the third method. Using the second method of measurement it could mean that each side of the mesh was 4 in.—not a very likely mesh size perhaps.

The only two unambiguous terms are the number of rows to the yard and the length of side of the mesh referred to as the "bar" and these are to be recommended. Throughout this work the number of rows to the yard will be used as this is the method of measurement used most frequently in the literature.

Selectivity

It has been found that different size meshes sample shoals in their own characteristic ways. Use of one mesh only, therefore, would give an inaccurate picture of the types of fish in a population, especially with regard to length, and therefore age, weight and sex.

Hickling (1939) did some experiments from the research ship *Onaway* using, besides nets with the commercial mesh size, also nets with meshes 42, 46 and 52 rows to the yard. Some of his results are given below.

1. *Length*

When plotted against the percentage frequency, length, over all the experiments, gave the trimodal curve shown in Fig. 8.3. The commercial net took fish mostly of 23 cm in

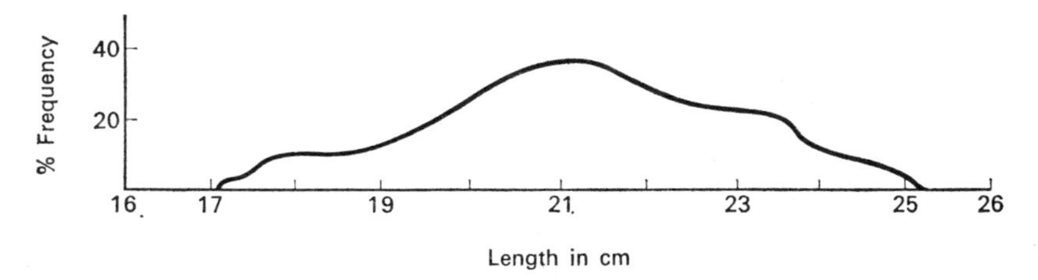

FIG. 8.3. Length–frequency composition of English pilchard population from different mesh nets—overall result.

length. The net with 42 meshes per yard took mostly 21 cm fish, the net with 46 rows per yard took mostly 21 cm fish, but captured many less fish than the 42 mesh net. Finally, the net with 52 meshes per yard captured very few fish, the largest proportion of which were 21 cm long. However, an almost equal number of fish of 18 cm and 23 cm length were caught. The largest ones caught by this small mesh were not gilled but were taken "by the nose" (Table 8).

TABLE 8. CHARACTERISTICS OF PILCHARDS CAUGHT BY DIFFERENT MESH SIZES

Fish caught	Size of mesh in number of rows to the yard			
	36	42	46	52
Length (cm)	23	21	21	21, 18, 23
Age (years)	6, 5, 7	5, 3, 4, 6	3, 5	
Weight (g)	101	78	73	
Number	621	1973		

2. *Age*

The commercial net caught mostly 6-year-old fish, but also many of 5 and 7 years old.

The net with 42 meshes per yard caught mostly 5-year-olds with many 3-, 4-, and 6-year-olds, and that with 46 meshes per yard caught mostly fish of 3- and 5-years-old (Table 8).

If only one type of net had been used, a completely erroneous impression of the age composition of the shoal would have been obtained. Furthermore, because a drift net tends to select fish of a certain length and age, the average length of each age group is affected. For example, the nets of 42 rows per yard took 5-year-old fish, which were of an average length of 21·2 cm, but 5-year-old fish taken by the commercial drift nets are of an average length of 21·9 cm. These two results taken out of context could be considered as providing strong evidence that there were two separate stocks involved which had different rates of growth. However, the pilchard has a protracted breeding season so that such differences of length with age might well be expected and consistent with a single population being selected by nets of different mesh size.

3. *Weight*

The average weights of fish caught in the three largest-meshed nets were as follows:

(a) The net with 36 rows/yard (commercial size) 101 g.
(b) The net with 42 rows/yard 78 g.
(c) The net with 46 rows/yard 73 g.

The larger-meshed nets took large fish in good condition, while the smaller meshes took a greater number of small fish in good condition plus larger fish in poor condition. On the whole, therefore, the mesh size would appear not to influence the overall condition of the fish caught as the two factors mentioned above tend to balance one another.

A most interesting point emerging from these experiments was that the net with 42 rows to the yard was the most productive throughout the trials. The three nets of this mesh took 1973 fish as against 621 fish in the three nets of commercial size. Thus, although the fish taken in the commercial nets were heavier by 23 g on average, the smaller-meshed net took 90,873 g of fish more. It would be worthwhile to investigate the effect that a change from fishing using nets with a mesh of 36 rows per yard to a mesh of 42 rows per yard might have on the pilchard population. Such a size net would tend to remove the 5-year-old fish, whereas now mostly 6- to 7-year-old fish are removed. This might have far-reaching effects on the breeding of the fish, as pilchards often do not become sexually mature until after they are 4 years old. However, experiments would show whether a change to such a mesh size would benefit the industry. Removal of younger fish with the smaller meshed net could lead to greater productivity in the population as a whole.

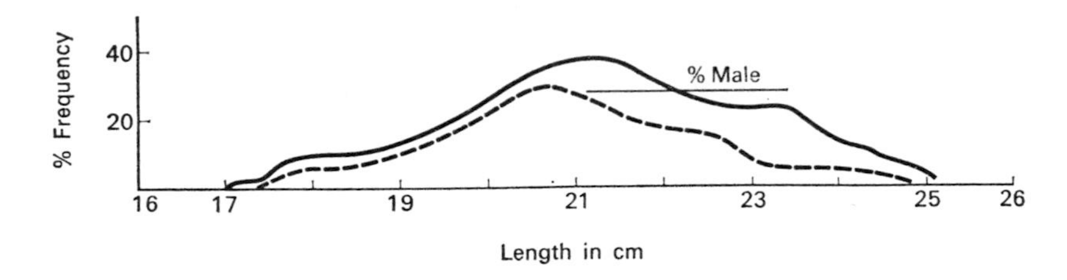

Fig. 8.4. Sex distribution of English pilchard population.

4. *Sex*

A surprising result of these experiments was that males seemed more abundant than females, forming 62 per cent of the total catch. In fact at lengths less than 20·5 cm they were in vastly greater proportion than the females (Fig. 8.4).

Other Methods of Catching the Pilchard

1. *Shore seining*

This was the method chiefly used in the past for catching pilchards, but its use gradually became less widespread. The numbers declined rapidly at the beginning of this century and the seine has not been seriously used for pilchards since the First World War. A full account of the method and the reasons for its decline have already been given.

2. *Trawling*

Trawling for pilchards was carried out during 1925 at Newlyn. The trawl had a well-corked headrope of about 40–50 ft, and a mesh throughout of 36 rows to the yard. The trawl doors were 4 ft by 3 ft. It was towed as fast as possible. However, neither the original trawl nor a later modification were successful, possibly because the horesepower of the vessel was insufficient.

3. *Mid-water trawling*

This was tried in 1954. Although it was reasonably successful at the first haul when 400 stones were caught, later attempts were not successful and the experiments were stopped. In 1961 and 1963 Bridger carried out some further experiments on a mid-water trawl from the R.V. *Madeline*. He had a standard echo-sounder and an Asdic echo ranger at his disposal. Unfortunately, the experiments were not a success. This was largely attributed to the presence in the water of *Noctiluca* during the summer which presumably provided enough light for the pilchards to avoid the trawl. In the winter shoals suitable for trawling were very rarely found (Bridger, 1965).

4. *Purse-seining*

(This is described in the section on the California fishery.) Purse-seining was used by Bridger experimentally (Bridger, 1965), but catches were not competitive with those of drifters working at the same time.

5. *Lampara netting*

Experimental fishing with a lampara net in conjunction with petrol lights hung over the side of the vessel to attract the fish have been conducted in the area with inconclusive results. The net, Spanish in origin, is rather like the ring-net (see below) but used by one boat, although a second, smaller boat is used when encircling the shoal.

6. *Ring netting*

This method of capture has been tried with results varying from complete failure to a single catch being as large as 2000 stones. It is only successful during the time of year that pilchards are really well shoaled. Two vessels are used, one of which shoots the net round the shoal. One warp of the net must first be attached to a buoy or to the second vessel (Fig. 8.5). Each vessel has a crew of five men. When shooting is finished both ends of the net are taken on to one vessel, and three men from the other vessel transfer to the hauling vessel. The messenger ropes are pulled in by a winch, but the headrope and net are pulled in by hand. Meanwhile the other vessel is situated on the opposite side of the hauling vessel and pulls the hauling vessel away from the net. This is for two reasons:

(i) the hauling vessel might otherwise capsize due to the weight of the fish in the net;
(ii) to prevent the hauling vessel from being dragged towards and into the net.

When the ends of the messenger ropes are reached, the winch is stopped and the partner vessel moves round and takes the headline of the net on board. The boats now have the bag of the net and its contents between them in the water. Long poles keep the two vessels apart whilst the catch is ladled from the bag by a mechanical dip net.

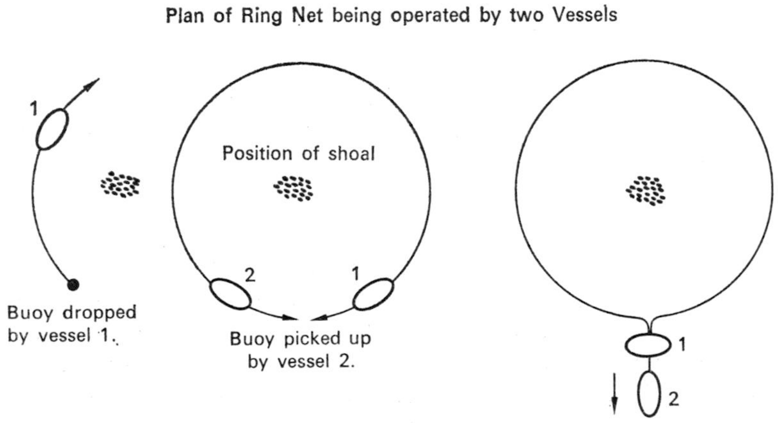

Plan of Ring Net being operated by two Vessels

Position of shoal

Buoy dropped by vessel 1.

Buoy picked up by vessel 2.

FIG. 8.5. Ring-netting operation by two vessels.

In this method of catching, shoals must first be definitely located. Visual natural signs can first be used, but these must be reinforced by further evidence, preferably from an echo-sounder. The shoals can never be caught during daylight by ring net as pilchards have very good sight and make an orderly exit from the end of the net which has not yet been fully closed.

A report by Hodgson and Richardson (1949) of some experiments on the Cornish Pilchard Fishery in 1947 and 1948 states that the pilchard can be attracted at night by lights, a shoal rising from 45 ft to 20 ft below the surface when searchlights were trained on it. Also they managed to attract a vast shoal to the surface by throwing pilchard meal overboard (a method similar to that used by Basque and Breton sardine fishermen who use salted cods' roe) although attempts to attract pilchard by salted cods' roe have since failed (Bridger, 1965). This shoal aggregated by the Hodgson and Richardson experiments when encircled was estimated at 2000 stones. An important point concerning ring netting is that shoals must be very accurately located otherwise they are left outside the net. A 200-yd ring-net when properly shot gives a circle of only 63 yd diameter. Certain chemicals have been used to help in location for easier shooting. For example, fluorescein has been poured on the surface of the water above the shoal to act as a marker.

Brief Comparison of Drift Netting with Ring Netting

Probably the chief advantage the ring netter has over the drifter is the much smaller time taken to operate the ring net compared with a fleet of drift nets. Leaving harbour after the drifter the ring netter can be back with a good catch whilst the drifter is still at the nets. Also when back in harbour there should be no residual fish to shake from the net as is the case with the drift net on many occasions. Thus the life of ring-net fishermen should be less arduous than that of the men on drifters.

Another advantage with ring netting is that the net is not shot until a sizeable shoal is indicated by the echo-sounder. Having found the fish, the ring-netter is usually certain of a good haul unless the net is badly shot. However, even if the drifters find good indications of a shoal and then shoot the nets, the fish may miss the nets almost completely, the ring netter is, therefore, dependent on fishing dense shoals of fish and does not make economically worthwhile trips unless such a dense shoal is located, whereas the drifter can pay its way even when the fish are dispersed.

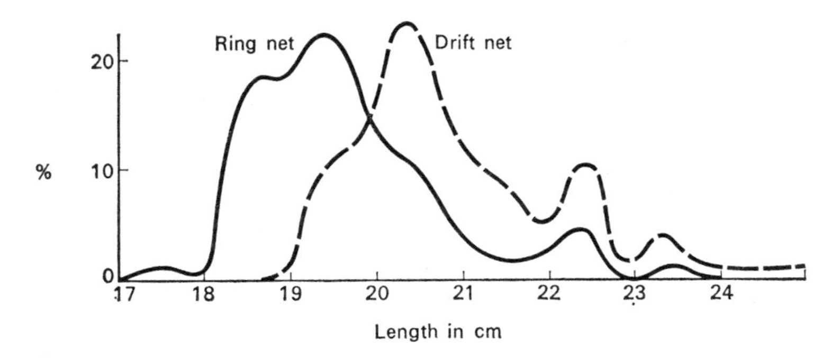

FIG. 8.6. Difference in length of pilchards taken by ring and drift nets.

The fish caught by ring nets should be in better condition than those caught by drift nets, as they are not gilled if the mesh is the correct size. The fish usually only lose a few scales but, unfortunately, they may get bruised.

It is a criticism frequently levelled against the ring net method that, as it samples the population randomly, the fish caught by this method are not so useful for the canners. The drift net, on the other hand, is selective for a certain size. However, if a graph comparing the percentage of each length caught by the two methods is drawn up it is found that they follow a similar pattern (Fig. 8.6). The ring net has about 18 per cent of its catch amongst a length group untouched by the drift net, namely 18·5 cm. However, the largest proportion of the catch of the ring net is between 18·5 cm and 19·5 cm. This should be a very useful size for the canners as the peak of the drift net curve is only at 20·5 cm. Minor adjustments to decapitating machines are all that would be necessary to accommodate the slight overall change in size. The canners would then have a sum of approximately 40 per cent of the ring-net catch to deal with at nearly equal lengths as against about 22 per cent of the drifters' catch.

The main disadvantage at the moment in ring netting is that although the catches may be large, they are erratic, being so dependent on the presence of dense shoals. The canneries cannot always deal with the wide fluctuations of catch size with which this method presents them. No convenient method of storing pilchards during a glut has yet been found, as chilled brine of the concentration used in the past hardens the scales so much that they will not come off in the descaling machine. New methods of dealing with pilchard gluts have been experimented with (see p. 82), but these are not used at all at present. The only method of dealing with gluts of pilchards at the moment is the old curing method, but since the war the demand for the salted material has been very low, although there are curing factories still operating in Cornwall.

Finally, in theory the ring netter is going to be kept in harbour more frequently by weather conditions than is the drifter. Nevertheless, in practice it was found, in the experiments referred to earlier, that weather which kept the ring netter in also tended to keep the drifters in.

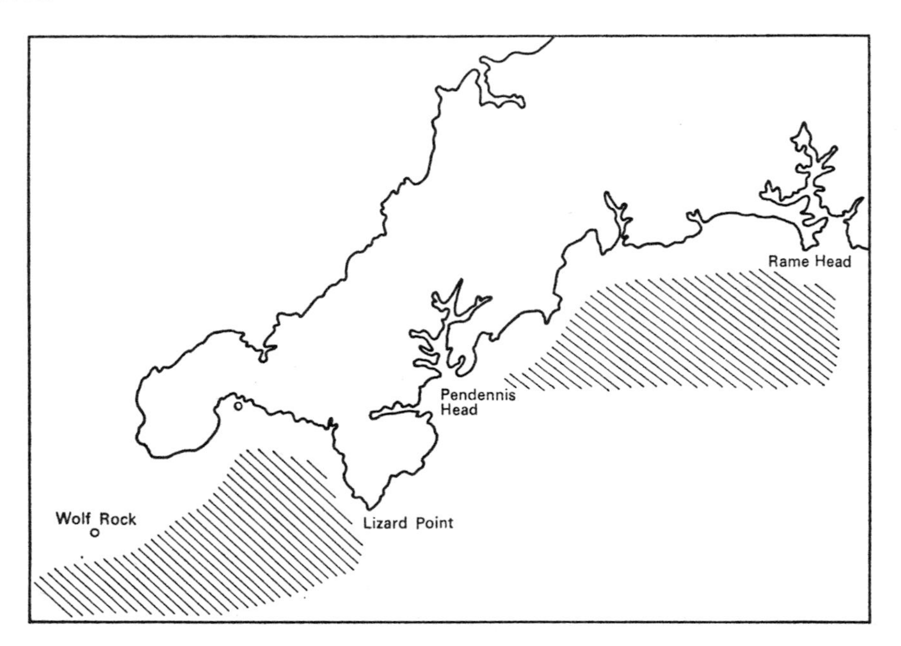

FIG. 8.7. Areas of pilchard drifting in Channel.

Conclusions

With the drifting method of catching and the type of vessel in use today, the fishermen are limited to regions within a few miles of the coast. The map (Fig. 8.7) shows the areas where pilchard drifting occurs. With the size of vessel which they use at the moment the Cornishmen are also wise not to leave harbour on stormy days.

With larger vessels using other methods of catching, such as the mid-water trawl, voyages further afield and with greater regularity could be undertaken.

There is no denying, however, that with the extension of the field new problems would arise. These are primarily connected with preservation of the catch. The pilchard deteriorates very rapidly after death, this being largely due to its being a plankton feeder as the food quickly liquifies. At the moment no method of preservation of the catch is used on pilchard drifters, not a single boat carrying ice. This is not because there are no ice-making facilities at the ports, as the trawlers in the area have regular supplies of ice.

The *Madeline* project mentioned earlier ended at the end of January 1963, and Bridger (1965) stated in his report that after trying many methods of fishing, drifting was the only method that would give consistently worth-while catches with vessels up to 100 ft in length; trawling might possibly be used by vessels above 100 ft in length and such vessels could probably make supplies of pilchards available in the winter as they would not be confined to grounds as close to the shore as the present drifter fleet.

CHAPTER 9

PROCESSING TECHNIQUES

THE people of the West Country have been catching pilchards for more than 400 years. The ancient method of curing the fish by salting produced a product which had a large market abroad, but which did not become well known in Great Britain outside Devon and Cornwall. The rapid development in canning techniques during the last 30 years has made the preserved pilchard an acceptable food for the rest of Britain and should have enabled the industry to expand. This has not happened; but it has allowed the industry to survive. Such is the importance of processing techniques, that the prosperity of the whole industry depends upon the acceptability of the article they produce.

Local salting of the fish has probably been done ever since the fish were first caught in bulk, and such a product was a very useful foodstuff to store for the winter. Probably most of the houses in the fishing villages around the Cornish coast had their own curing cellars. The cured product was once an important export to Italy, but, as we shall see (Chapter 11), this has declined since the First World War. However, from the eighteenth century to 1913 the quantity of salted pilchards exported was quite considerable. Although the amount fluctuated from year to year it was about 4000 tons as an annual average. Before 1800 Italy was buying this product, as shown by some figures published by a Dr. William Borlase in 1758 (Couch, 1835). The average quantity for the 10-year period from 1747 through 1756 was 29,795 hogsheads, which is about 6000 tons. Even before this, in the late sixteenth and early seventeenth centuries 95 per cent of the production of salted pilchards was being exported as the following quotation shows, although the distribution of the exports was not mentioned.

> This commodity at first carried a very low price, and served for the inhabitants' cheapest provision; but of late times, the dear sale beyond the seas both so increased the number of takers, and the takers jarring and brawling one with another, and foreclosing the fishes taking their kind within harbour, so decreased the number of the taken, as the price daily extendeth to a higher rate, equalling the proportion of other fish: a matter which yet I reckon not prejudicial to the commonwealth, seeing there is store sufficient of other victuals, and that of these a twentieth part will serve the country's need, and the other nineteen pass into foreign realms with a gainful utterance. [From Carew (n.d.), the de Dunstanville edition, 1811.]

It was not possible to preserve food using methods other than salting until a Frenchman, Appert (Malcolm-Smith, 1949), devised a method, in 1810, whereby food could be preserved for some time in boiled water. Canning developed in America at the end of the nineteenth century (Cutting, 1955), and by 1918 there were thirty-two canneries in California alone which handled nearly 70,000 tons of sardines between them. In Great Britain the canning industry was important before 1914, but the canning of fish did not develop until the inter-war period. Even by 1935, the total production of canned fish in Great Britain had only reached 9000 tons.

In the past there have been several canning factories in the south-west, some of which dealt exclusively with pilchards, at present the number is considerably reduced. The methods of curing and canning pilchards are described below.

Curing

The curing of pilchards used to be very much more widely carried on than it is at present. In June 1963 there were three curing factories in Cornwall. Of these, two were at Mevagissey and one at Newlyn. However, one of the Mevagissey curers and the Newlyn curers were the same firm which has since closed down. At the present time there are two small firms curing pilchards at Newlyn employing only eight full-time and ten part-time staff between them (Great Britain, 1966).

As mentioned earlier, curing also used to be carried out domestically. Writing in 1883, Wilcocks says: "Fishermen in Cornwall always retain a stock of salt pilchards at home for the use of their households through the winter, not of course that they live exclusively on this diet, but they always like to have them at hand, and if they get a good crop of potatoes also, they look forward to the winter without anxiety."

Commercial curing tanks are usually placed as near the shore as possible. They are from 10 to 15 ft square and about 10 ft deep. Broadly, the process is as follows. The fish are first stacked in salted layers by the sides of the tanks, in small heaps. These piles are then shovelled into the tanks where they remain for at least 3 weeks. During this time a strong brine should be produced, a stronger brine effecting a better cure.

After the time in the tank the fish are washed and packed radially in cylindrical wooden containers. The contents of the containers are then pressed, causing oil to run from them. This is collected, and has various uses, some having been taken at least in past time by British Railways for lubricating purposes (Kennea, 1957). At one time it was used by curriers in preparation of leather. After the initial pressing more fish are added, then pressure is applied again and the process repeated until the container will hold no more fish.

The casks are in three sizes:

1. the half cask, holding 120 lb of fish,
2. the quarter cask, holding 60 lb of fish, and
3. the one-eighth cask, holding 30 lb of fish.

The full cask which held 240 lb is no longer in use. It seems that such casks were similar in design but smaller than the old measure of fish known as the hogshead. They were made of open staves, and were shut by a lid known as a buckler to which the pressure was applied by means of a weighted lever. The oil would then run out of the openings between the staves of the cask which used to produce about 2 gal of oil.

As mentioned previously, the number of curing factories working at present is very much below past figures, having more than halved in the 11 years from 1955 through 1965, and has in fact never recovered properly since the First World War. Italy at one time was a large importer of the cured product, which was cheap yet highly nutritious. However, the standard of living there has risen in the past 40 years, and substantially so since the last war, so that fish treated in this manner may perhaps be less acceptable, although no major post-war effort has been made to reopen the market.

Although the Italian market is at present at a low level for cured pilchards, there is still a largely untapped market in Eastern countries and this is probably worth while investigating.

Canning

At the moment the largest percentage of the catch is canned. Fish caught on one night and landed at any time from midnight to the early hours of the morning cannot be canned before the factory starts work for the day. Rarely, on occasions when a glut has occurred, canning might be continued through the night. During the time before canning takes place, nothing is done to help to preserve the fish. Ice is not used on board pilchard drifters, nor are the fish covered by ice or held in chilled brine after landing. Also, chilled brine hardens the scales so much that they do not then come off in the descaling machine, although this medium of temporary refrigeration had been successfully used in California (Davis *et al.*, 1945). However, the fish could be held in ice to help maintain their condition during this waiting period. In some experiments done in connection with the South African pilchard industry, it was found that 270 lb of ice enabled 1 ton of pilchards just floating in brine to be chilled from 14° to 4·4°C in a wooden holding tank in 1½ hr. It was found that more uniform results were obtained, but temperatures rose slightly above 4·4°C when more than the minimum amount of sea water required to float the fish was added to the tank. If fish were chilled too much below −7°C and were then put through the decapitating machine, the accompanying degutting process was not always complete, small parts of the gut being left in the body cavity. The percentage of fish affected in this way increased with lowering of temperature thus (Cooper, 1957):

At 4·4°C 2 per cent were not completely degutted,
1·6°C 10 per cent were not completely degutted,
0·5°C 48 per cent were not completely degutted.

Refrigerated sea water could probably be used for storage on shore (Roach *et al.*, 1961), although the size of the vessels at present used is too small for a R.S.W. plant actually to be installed on them. It does seem, therefore, that a method of holding pilchards temporarily could be found quite easily. Methods of keeping fish in storage for longer periods have been investigated, and these will be dealt with in Chapter 15.

An outline of the canning process is given below. Although variations in procedure are found from one factory to another the principles are broadly the same (Fig. 9.1).

1. A preliminary inspection usually takes place in which damaged fish and those considered too soft for canning are removed.
2. The fish are then decapitated by a machine which also pulls out the gut, leaving the gonads *in situ*. The machine has steel slots on a belt operated system which can deal with sixty-five to eighty fish per minute. The fish should not be split open by this process if they are in good condition. Offal is discharged separately and the fish transferred to a descaling machine.
3. The descaling machine has a cylinder on which are fitted some revolving ribs. Two plough-shaped baffles in the centre of the machine cause the fish to rub against the ribs and against each other, thus loosening the scales (Dewberry, 1957). The fish leave this machine and:
4. are then soaked for 10–20 min in tanks containing brine. The brine should contain no magnesium or calcium sulphates as these salts may adversely affect the colour and texture of the fish (Jarvis, 1950).
5. The fish are sometimes then gently washed in water.
6. Next they are transferred to the canning section of the factory. Cans are placed on a machine which delivers a regulated quantity of tomato sauce into them.

7. From this machine the cans pass on to a belt along which women workers are ranged. They place the fish in the cans, tail ends alternating with the head ends. (There are three fish in a 7-oz tin.)
8. At the end of this belt the lid is placed in position and is partly closed in two places.
9. The cans then pass through a steamer in which they reach 210°F, after which they are fed into another machine.
10. This stamps the lid finally into position.
11. They are placed in wire baskets and transferred to an oven which operates at a temperature of about 240°F. This softens bones and destroys any spoilage organisms. Seven-ounce cans stay in the oven for 50 min. One oven can take 1500 tins.
12. On removal from the oven they are sprayed with cold water to cool them.
13. They are labelled and packed in boxes and then, finally:
14. Left to mature for 30 days before dispatch.

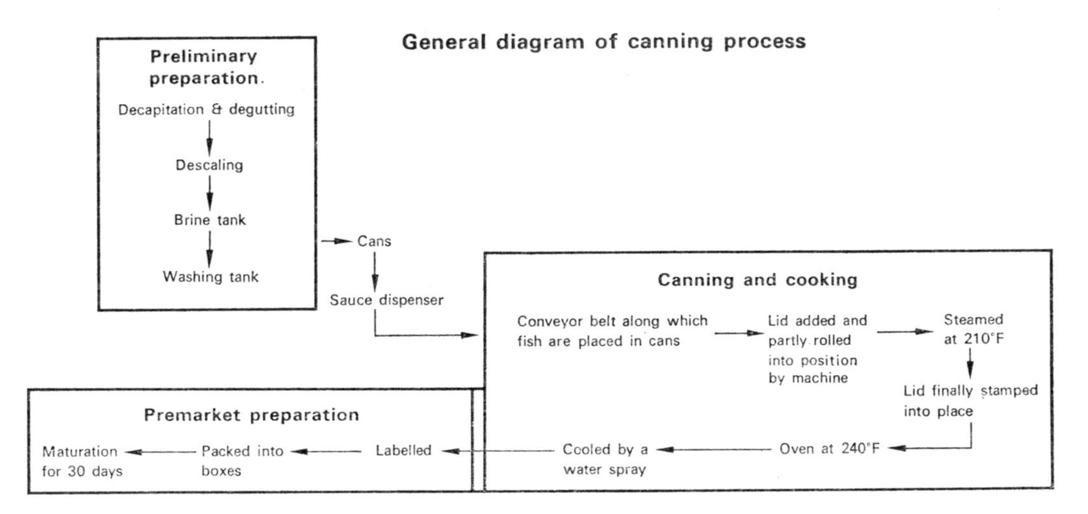

FIG. 9.1. Flow diagram of canning factory.

If the account of the canning procedure for pilchards is compared with that for related species, such as herring, it is found that there is only one main difference. It was mentioned above that, after being placed in the tins the fish passed through a steamer at about 210°F. This is the pre-cooking process which tends to make the flesh firmer and yet produce the minimum amount of liquid in the can. This is the part of the process which is not included in canning herrings (Hughes, 1957). Although the non-inclusion of this stage reduces the amount of machinery required, it would seem that the product is less satisfactory.

The period of 30 days' maturation is recommended for other canned fish as well as pilchards. During this time the flesh becomes firmer, a process which may continue for more than 1 month, and changes occur in the flavour of the contents. It is thought that characteristic flavour and odours of fish are imparted by certain fat-soluble substances present in the flesh in small quantities. These substances are of low molecular weight, but, apart from this, very little is known about them. Other substances present in larger quantities, for example the water-soluble trimethylamine oxide, also influence the flavour of the product. Slow decomposition of this substance is known to occur in canned sardines, so it is likely that this is another of the chemical processes involved in imparting to products their particular flavours (Hughes, 1957), and is a partial explanation, at least, of why the period of maturation is necessary.

One firm in Newlyn is able to produce 2800 cans of pilchards per hour. The cost of the cans themselves varies according to size, and whether they are lacquered inside or not. This firm deals largely with 7-oz oval, 14-oz oval and 8-oz round cans. The purpose of lacquering the cans is to provide an impermeable barrier between the can and its contents. Some lacquers include a pigment, the purpose of which is to adsorb any sulphur compounds, thus preventing the dark staining which occurs in some canned foodstuffs. (Leonard Hill (Books) Ltd., 1957.)

Fuel consumption of the factory naturally varies with output but, when working every day, is in the region of 3 tons of coal per week.

Labour has to be called in at the beginning of the fishing season, and is employed for the season. Extra women are brought in if it seems likely that the intake of fish has increased to a new steady value. Casual labour is not employed. All staff are laid off during the closed season by this particular firm as they deal exclusively with pilchards.

It is possible that in the future, factories will be able to utilize pilchards in ways other than the usual forms of canned in tomato sauce and fish paste. For instance they could be canned in olive oil, which would be preferable for some tastes. There are many different sauce flavours used to make the herrings canned in Germany more attractive (Sidaway (n.d.)). There is apparently a state of healthy competition between different canning factories to see which can produce the most interesting and the most flavoursome sauces. Some of the flavours produced by the Germans include lemon, curry, tartar, mushroom, crab, shrimp, mustard, mayonnaise, red wine, white wine and beer. The mixing of the sauces is often done by the manager of the factory, in a room set apart for the purpose. An egg substitute, the composition of which is kept secret, is used as a stabilizer in many of the sauces, which are electrically stirred and then further stabilized by the use of a high-speed centrifugal homogenizer. No exceptionally expensive apparatus is used for dispensing the sauces. As the sauces are usually mixed in a room above the general factory level, they can be piped gravitationally to the point where they are measured into the cans. The cans have half the sauce (50 g) placed in them before the fish are added, and the other half afterwards. This dispensing is done either by hand, using a ladle, or by a simple valved nozzle of the sort used for petrol.

As the pilchard is so closely related to the herring there does not appear to be any significant biological or chemical reason why it could not be presented in a similar variety of sauces. The expense involved in new apparatus would not be great, but the improvement in taste, variety and consumer appeal would be considerable.

Another interesting and tasty product could be made by marinating pilchards. One manager of a canning factory who was consulted concerning this thought that it might be rather a costly product, slow to make. However, marinated pilchards are very popular in the West Country and with suitable advertising might be successful in other parts of the country.

Mechanization

Most of the canning procedure is now mechanized, but there are parts of the process which could be further mechanized. For example, in some factories the area where the fish are prepared for the cans is kept separated from the actual canning area. The wire baskets containing the fish are lifted by hand onto trollies and wheeled to the canning area by hand labour. From the economic point of view this seems a wasteful process. The fish could be transferred from the brine, or the washing process if this is included, by an automatic device. From here they could pass into a chute, or on to a conveyor belt for transference to the canning area.

There are two main elements involved in a decision to mechanize the process. The first is a question of costs. The machinery necessary would obviously take a certain time to pay for itself, but at the present rates of pay a machine costing a few hundred pounds could probably be amortized within a year or so. The second point to consider is the time factor. Apart from the immediate saving of labour, the whole factory might be able to run more efficiently and smoothly if the canning conveyor belt were supplied continuously with fish.

Once at the canning area of the factory the fish are placed in the cans by hand. It is probable that no satisfactory machine has yet been specifically designed for the purpose of feeding fish into cans. However, if it were possible to make a change to mechanization of this process (and it is likely that a machine designed for canning some other foodstuff could be adapted to this purpose), then there are several advantages which would be gained by such a change.

Normally about six women are employed for this purpose alone. A machine that could dispense with employment of these people would be a particular boon in the West Country as it would be able to prevent the recurrence of the perennial problem of recruitment of this female labour. This is becoming more acute, and will continue to do so, now that the tourist trade is so flourishing in the West Country. More part-time jobs connected with tourism have become available for women and these are more lucrative and offer more pleasant working conditions than a job in a pilchard canning factory. On the rare occasions when gluts of fish occur, a machine (which may or may not be of variable speed and/or capacity) can be worked throughout the night without entailing a disproportionate increase in expenditure which is inherent when human labour is necessary, due to the overtime rates.

The only other major part of the process of canning which remains unmechanized is again a transference. After steaming, the cans are fully sealed and packed into wire baskets. Both this packing of the baskets and their consequent removal to the ovens are done by hand. It is probable that this stage will remain unmechanized unless the design of the ovens is basically altered. If, when the time comes to replace the existing ovens, a new design has been evolved incorporating possibilities of mechanical loading and unloading, then it would be advantageous to mechanize this stage also.

Further mechanization will certainly come to the pilchard canning industry just as in other industries. Another universal trend in mechanization will also affect the industry eventually, that is the incorporation in one machine of processes which were formerly done by several. Just as in other branches of the pilchard industry, however, changes from accepted methods to newer, more economical methods will be slow in taking place. This is partly because the capital necessary for the changes is apparently not available, partly due to the innate conservatism of the men responsible for such decisions, and partly due to an inherent danger in the area of unemployment.

CHAPTER 10

SEASONAL CHANGES

IT WAS established in Chapter 4 that the weight of the pilchard and the composition of its flesh vary quite considerably over the whole year. This is due to several causes, the first of which is that there is variation in the food taken. It was noted that the usual diet is the fairly oil-rich zooplankton, but phytoplankton is also an important item of the diet and this is less oily. There is bound to be some reflection of these differences in the flesh itself. Secondly, the actual amount eaten and the quantity assimilated by the flesh and various organs is different at different times of the year. It has in fact been shown that various regular seasonal differences in the composition of either the flesh or the organs occur in the pilchard of the Channel.

Thus, Hickling (1945) found that the pilchard stock off the Cornish coast undergoes a seasonal cycle of weight changes. In studying these weight changes he divided the fish into three groups according to length. These were:

(a) small fish, not greater than 20 cm long,
(b) medium fish from 20 to 23 cm long,
(c) large fish, not less than 23 cm long.

He then took fish whose lengths were representative of each group; those of 19 cm length representing group (a), those of 21 cm length representing group (b) and those of 24 cm length representing group (c). The weights of these fish were then plotted as their monthly mean weights after they had been gutted. The graph shown in Fig. 10.1 gives the results of this work for all three classes of fish from July 1935 to November 1938 inclusive.

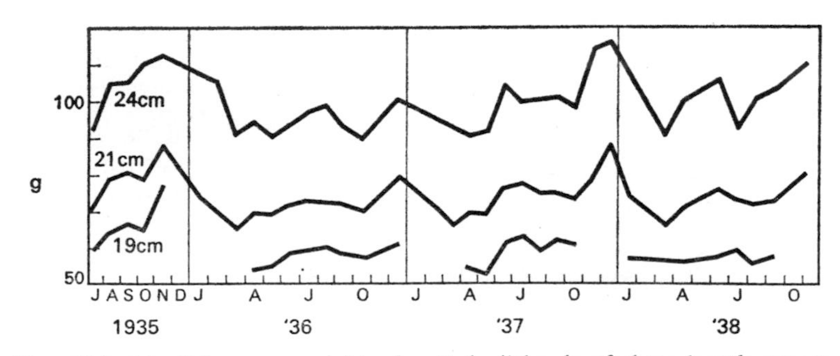

FIG. 10.1. Monthly mean weight of gutted pilchards of three length groups.

It can be seen from this graph that all three size groups are affected in a very similar way. Although the figures for the smallest length group are not complete, those that are available show the same trends as the other two groups. Broadly, the cycle, consisting of two minima and two maxima, has the following pattern. There is a minimum weight which occurs, usually in March, but may be as late as April. There follows a fairly rapid rise in weight to the first maximum which is in June, July or August. After this the weight of the fish falls, usually quite gradually, but there may be a rapid decline, to the second minimum. They sometimes lose weight so much at this time that they reach their weight of the previous March. From here the weight rises to the second maximum which is higher than the first and occurs in November or December.

Such changes in weight are intimately connected with the life of the pilchard, especially the feeding and breeding cycles. It was seen in an earlier chapter that the feeding cycle is closely related to the availability of plankton, and it was pointed out that there was a high rate of feeding between April and the end of July with a maximum reached in June. From July until mid-September the amount of food taken was seen to decline rapidly. Following this there was a further feeding peak about mid-October which was occasioned by the autumnal plankton outburst. From November to February it was found that very little food was taken and thus the fish had to live on their resources during this time. This is why, in March or April, they are found to have an extremely low weight, and are ready to start feeding again as soon as the spring brings an outbreak of planktonic food.

This cyclical pattern of weight changes appears to occur regularly in the pilchards found in the Channel. However, as would perhaps be expected, the changes seem to be largely induced by the environment, as Ramalho (1933, 1935) working with the same species off Portugal found that the seasonal cycle of weight changes included only one maximum. What is unexpected is that this maximum could apparently occur either in June or August or in November or December in different years.

It will now be shown that the overall weight changes are affected by other factors. Of these factors, two fluctuate considerably throughout the seasons and one varies only slightly during the year. These factors are:

(i) Fat content.
(ii) Water content.
(iii) Protein content.

There is also a fourth factor making up the total percentage composition of the fish which will be mentioned later. These factors will now be considered in turn.

(i) *Fat Content*

The broken line in the graph given in Fig. 10.2 represents the variation of the fat content of the flesh of medium sized pilchards, group (b). This curve was obtained by plotting the

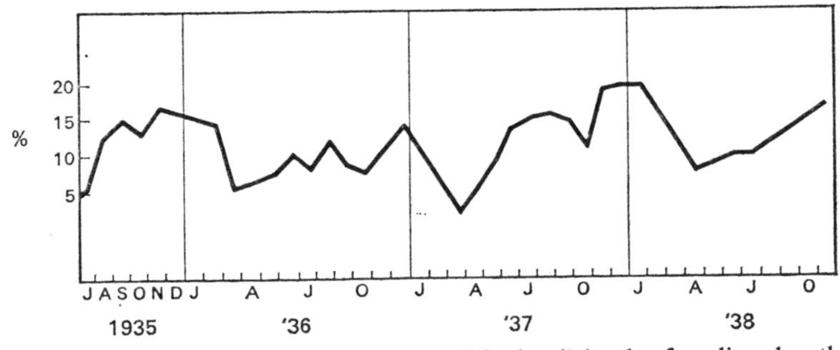

FIG. 10.2. Monthly mean percentage of fat in pilchards of medium length.

monthly mean weight of fat expressed as a percentage of the mean weight. The mean weight of fish of 21 cm indicated in Fig. 10.1 appears to be very closely correlated with the fat content. The latter is shown to rise from a minimum in March or April to a peak in August. There is then a decline, as there is in the weight curve, which reaches its lowest point in October. This is followed by a further rise to a maximum, higher than the summer one, in November or December. Because of the close correlation between these two factors, it seems certain that seasonal fluctuations in fat content govern the seasonal changes in weight.

Fat, besides being stored in the flesh, is also stored in the mesentery and liver. The amount stored in the mesentery is considerable, and during the months of December to February or March is drawn upon to a greater extent than the body fat. It is, therefore, very important to the animal during this time, although it is slower to regenerate than the body fat. The regeneration starts in May or June. Maximum mesenteric fat is first reached in August or September after which there is only a slight decline. However, the store is replenished again by

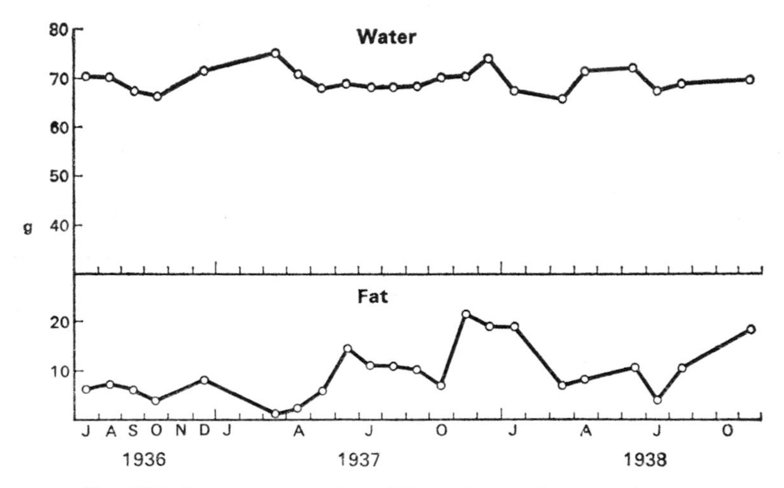

FIG. 10.3. Inverse correlation of fat and water in large pilchards.

November or December. The liver of the pilchard is very small and the quantity of fat stored in it is not important in the cyclical changes being considered.

(ii) *Water Content*

Hickling (1945) found that the quantity of water in the flesh of the pilchard varied inversely with the fat content. He stated that the inverse correlation which exists between the two could hardly be bettered in biological material. This is shown in the graph, Fig. 10.3, for pilchards of 24 cm length, group (c). Although the graphs for the two shorter length groups are not shown, it was found again that they all followed a similar pattern. There is not, however, the regular seasonal change that was found when the fat content was being considered. For example, from January to March 1937 the water content of pilchards 24 cm in length rose, but during the same period in 1938 there was a fall in water content.

(iii) *Protein Content*

This was calculated as a residual percentage after estimation of the other three factors. It was found that a slight seasonal variation occurred which followed the general pattern of the variation in fat content. This rather flattened seasonal cycle is shown in graphical form in Fig. 10.4. Hickling thought that the amount which the protein is drawn upon is greater in larger than in smaller fish.

(iv) *Ash Content*

This is the fourth factor mentioned earlier. The figures are obtained by combustion of the fish, and because of this method of obtaining them they may not be as accurate as the figures for the other three factors, due to a certain amount of boiling fat which is lost.

The composition of the pilchard at different seasons as judged by these four factors is shown in Tables 9–11. Table 9 gives the average percentage composition of the medium and large fish during March and April, that is, at the time of minimum mean weight.

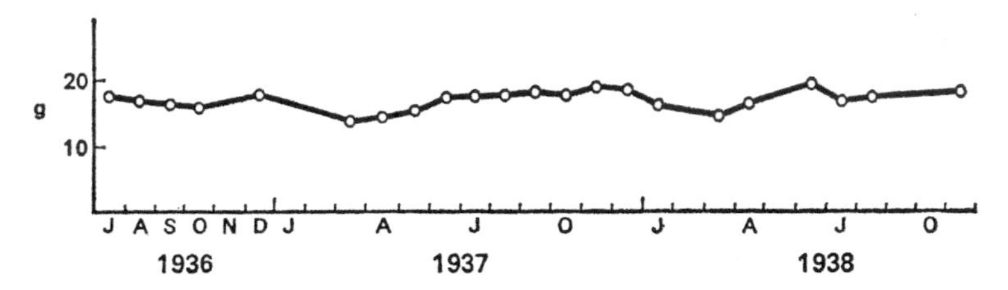

FIG. 10.4. Graph of protein content.

TABLE 9. AVERAGE PERCENTAGE COMPOSITION OF PILCHARDS AT THE END OF WINTER

Length (cm)	Fat	Water	Ash	Residue
20–22	5·8	75·0	1·9	17·2
More than 23	5·1	75·9	2·0	17·0

Table 10 shows the average percentage composition of all three size groups of fish which is approximately that at the height of the fishery in July and August.

TABLE 10. AVERAGE PERCENTAGE COMPOSITION OF PILCHARDS AT THE HEIGHT OF THE FISHERY

Length (cm)	Fat	Water	Ash	Residue
Less than 20	13·6	66·8	1·6	18·0
20–22	11·2	68·6	1·7	18·5
More than 23	9·4	70·2	1·8	18·6

The average composition of the two largest classes of fish during November, December and January, that is the period embracing the time when they reach their second maximum in the weight cycle, is given in Table 11.

TABLE 11. AVERAGE PERCENTAGE COMPOSITION AT TIME OF SECOND MAXIMUM IN WEIGHT CYCLE

Length (cm)	Fat	Water	Ash	Residue
20–22	17·2	63·6	1·6	17·6
More than 23	16·0	64·5	1·7	17·8

From these three tables it can be seen that the percentages of ash and residue remain comparatively stable throughout the year, but that at the height of the fishing season the percentage of protein, as represented by the residue, is higher than at other times of the year. As would be expected, the percentage of fat is higher, and consequently that of water lower, at times of active feeding, than during the period March to April after the prolonged period of little or no feeding.

What does become clear from a study of these figures is that at all times of the year the smaller fish are the most valuable from a nutritional point of view, taking fat percentage as being the most important single criterion by which to judge this. For example, at the height of the fishery a small fish of less than 20 cm length contains an average of 13·6 per cent fat compared with the average of 9·4 per cent fat for fish longer than 23 cm. Unfortunately, this property of high fat content does not coincide with the most favourable conditions of the fish for canning.

Suitability for Canning

The composition of the pilchard during the months now favoured as the principal fishing season is not as good from the nutritional point of view as during the months of November, December and January. It has already been pointed out that November, December and sometimes also January were at one time the most important 3 months of the fishing year. At this time the fish are at the peak in their weight curves and contain the highest percentage of fat. From the purely nutritional aspect, therefore, it would apparently be advantageous, once more, to increase the catch during the last 2 months of the year, rather than during the summer. Added to this is the fact that probably the summer catches contain a larger proportion of breeding fish than other catches. Fish which have been spawning tend to can badly and become "mushy", that is, are inclined to break up on turning out of the can or when lifted (Dreosti *et al.*, 1957).

Mushiness in pilchards was studied by the Cape Town Fishing Industry Research Institute and they considered that it may be caused by several factors, of which the following three are probably the most important. (It is probable that the factors mentioned below do not operate singly.)

(i) A physiological and physical condition of the fish associated with the spawning period.
(ii) Infection with the protozoal parasite *Chloromyxum thyrsites.*
(iii) Deterioration of the fish during prolonged storage.

Spawning fish are fairly easily detected by the fact that they are generally lean and lacking the layer of subcutaneous fat found in pilchards in good condition. Furthermore, they pass out of rigor mortis in 3–6 hr, as against the usual 8–12 hr. It has been suggested (Dreosti *et al.*, 1957) that spawning fish should not be canned unless they have been caught less than 8 hr previously. The information just quoted was specifically with reference to the condition of spawning pilchards of the South African stock, but it is probably equally true for Cornish pilchards. Whether the Cornish stock is infected by the protozoan parasite is not known, but this ought to be investigated in the near future.

At the present time, however, although pilchards are canned throughout the year with the exception of two short periods, the two times that the factories tend to close down are within the times when the fish are likely to be in a condition least favourable for canning. The first period is during February and March. Catches at this time are low because the boats are often prevented from going out due to unfavourable weather conditions and the fish are, in any case, not readily available. At this time the fish are at their lowest weight for any given size. From the nutritional point of view also the product would be of the least value at this time of the year as the fish are at the end of, or nearly at the end of, their period of fasting. The percentage of fat for any given length of fish is, during February and March, less than half its value at other times of the year, and less than a third of the value at the time of the second maximum (see Tables 11 to 13).

The other time that factories close is for their annual holiday, and this varies somewhat from firm to firm. For example, one of the chief firms to can pilchards closes for a fortnight at the end of July and the beginning of August. This is also a reasonable time to stop as it is likely that during this time a large proportion of the stock will be spawning. It is, however, rather unfortunate that the trend in increasing summer catches, noted in Chapter 7 will probably lead to an increasing proportion of spawning fish being canned, which in turn could lead to a deterioration in the standard of the final product.

CHAPTER 11

THE CHIEF OUTLETS FOR THE PILCHARD

ONE of the main complaints of the pilchard fishermen is that the number of markets open to them is very small, and therefore their catch is limited by the resulting lack of demand. In fact, if we investigate the markets open to the pilchard in either fresh or processed condition, at home or abroad, we find that the number is extremely limited and the fishermen's complaint seems justified.

There are four markets available to the pilchard fishermen, and these are:

1. the canning factories,
2. the curing plants,
3. those fishermen requiring bait, and
4. the fresh fish markets.

Each of these will now be dealt with in more detail.

1. From 1949 up to, and including 1956, except for 2 years (1952 and 1956), less than 50 per cent of the catch was canned (Great Britain, 1949–56). Since 1956, that is starting with 1957, more than 50 per cent of the catch has been canned. This is shown in Table 12 for the years 1949 through 1966.

TABLE 12. QUANTITY OF PILCHARDS CANNED FROM 1949
(Figures in tons)

Year	Catch	Quantity canned	% canned	Year	Catch	Quantity canned	% canned
1949	3020	900	29·8	1958	4148	3447	83·1
1950	3935	1730	44·0	1959	3617	1840	51·0
1951	5805	1900	32·7	1960	2986	1460	48·9
1952	4053	2640	56·7	1961	2700	1230	45·6
1953	5456	2300	42·2	1962		900	
1954	3001	1300	43·3	1963		500	
1955	4850	2500	51·5	1964		900	
1956	5946	2700	45·5	1965		N.A.	
1957	2957	2100	71·1	1966		700	

N.A. Figures not available.

It seems that the canning factories are now the chief buyers of pilchards. There were, until 1962, four canning factories in the area. However, since the closing of the one at Looe in that year, the following are the only remaining firms concerned with canning pilchards in the West Country.

1. Cornish Canners at Newlyn.
2. Duchy Products at Mevagissey.
3. Shippams Limited at Newlyn.

The last-mentioned firm has its canning section at Chichester in Sussex, but uses fish from Newlyn after they have been gutted and descaled there. Henry Sutton Limited, of Yarmouth is another important firm which cans pilchards. The fish are transported from the West Country heavily iced, in fast overnight lorries (Henry Sutton, personal communication, 16 February 1963).

One of these firms deals with a large percentage of the Cornish catch, which goes for canning, the other firms take varying quantities of the catch which remains after the other small outlets have been supplied. The largest firm canning pilchards offers about 2*s.* 9*d.* per stone for unboxed fish. However, they have started supplying boxes to the fishermen, asking them to pack the fish in these as soon as they are shaken out of the nets. For boxed fish up to 6*d.* a stone more is paid. The number of boxes for each boat is limited by the size of the crew as, at the moment, they are issued five per crew member. The idea is still in the experimental stage, but it can already be seen that the fish which are boxed immediately after catching are in much better condition than the unboxed ones. This is because there is much less weight on them than in the usual fish hold so that they remain uncrushed. The capacity of these steel boxes is 6 stones and the number issued in each boat could be increased; the extra price paid to the fishermen pays off from the canners' point of view as fish of superior quality are obtained, and from the fisherman's point of view makes the extra effort involved in packing them worthwhile. Some fishermen tried shovelling the fish into these boxes on arrival back at the port, thus attempting to obtain the extra price per stone. However, the improvement in quality of fish boxed straight from the nets is great enough to make this practice detectable.

2. The next market open to the pilchard fishermen is that of the curers. This industry once formed an important part of the economy of the whole pilchard industry. There have been exports of salted pilchards from England to Italy at least since the sixteenth century. However, since the First World War this market has never fully recovered, although it has kept going at a low ebb until the present time. By 1955, however, the exports had dropped to 100 tons and since then they have not been recorded in the statistics of the Food and Agricultural Organization of the United Nations. The amount of the catch which was cured from 1949 to 1955 is shown in Table 13. This quantity is also given as a percentage of the total catch.

TABLE 13. QUANTITY OF PILCHARDS CURED
(Figures in tons, F.A.O., 1955, 1960)

Year	Catch	Quantity cured	% cured
1949	3020	1400	46·4
1950	3935	700	17·8
1951	5805	800	13·7
1952	4053	700	17·2
1953	5456	700	12·8
1954	3001	500	16·7
1955	4850	100	2·0

Estimates of the quantity cured in the years from 1956 onwards have been made in two ways. The first method was suggested to me by the White Fish Authority, and was according to the following formula:

"Total catch — quantity canned × 3/2 = total wet weight cured.
Total wet weight cured — 20 per cent to allow for loss of weight during curing = total quantity cured."

The second method was from figures kindly supplied to me by the main salting firm of Cornwall. They told me the quantities of fresh pilchards that they had bought for curing, from 1956 onwards. These figures were reduced, on their recommendation, by 27 per cent to give the amount cured. This allows for loss of weight during the curing process and also for fish which were unsuitable for curing which had been included in the fresh, wet weight. This gave figures for the total production of this particular firm which is probably 75 per cent of the total production of this commodity for the whole country. Increasing the totals for this formula by 25 per cent gave the overall total for the whole country. The figures arrived at by these two methods are compared in Table 14.

TABLE 14. RESULTS FROM TWO METHODS OF ESTIMATING CURED PRODUCTION OF SALTED PILCHARD IN GREAT BRITAIN
(Figures in tons)

Year	Official estimate	"Private" estimate	Year	Official estimate	"Private" estimate
1956	974	89	1961	440	167
1957	271	107	1962		196
1958	210	134	1963		
1959	233	264	1964		
1960	458	206	1965		

It can be seen that the two figures do not tally, possibly because the official estimate is dependent on the total catch which is an unrealistic base for calculations at the present time. However, it is possible to conclude from these figures that as an export product cured pilchards have become insignificant.

From Table 13 it can be seen that although the quantity cured in 1950 dropped by half when compared with the previous year, the percentage fell by more than two and a half times. Since then the percentage of the catch cured has stayed below 20 per cent and in 1955 dropped as low as 2 per cent.

In comparing the two markets, that for canning and that for curing, since 1951, the actual quantity canned has not altered a great deal. In general, however, between 1951 and 1958 there was a tendency for the percentage of the catch which was canned to rise, but since 1958 the actual amount canned and the percentage of the catch canned have been falling again. In 1957, 1960 and 1961 the total catches dropped below 3000 tons and the absolute amount of pilchards canned was very small. For cured pilchards, the quantity treated fell consistently from 1951 to 1955. During 1954, when the total catch was very low the percentage of it cured rose slightly, but this did not reflect an absolute rise in the amount cured. Since 1956, the first year during which the quantity of pilchards cured was not recorded, the estimates of the quantities cured have all been above the 1955 figure (except for the 1956 "Private" estimate), but quite negligible in terms of an export product.

3. The next market for pilchards is the portion sold as fresh fish to fish shops and hawkers. This part of the trade is entirely local. As a "home-produced" fish the pilchard is eaten to a small extent in the West Country, but beyond this it is hardly known as a fresh fish at all. It is said that a certain amount was at one time transported to inland markets such as Billingsgate and Birmingham. Although this quantity may have been considerable (Wilcocks, 1883), it is now very small and no inland markets sell pilchards regularly and no statistics are available on the subject.

4. The only other type of demand for pilchards comes from those fishermen who use these fish for bait. From 4 to 8 per cent of the catch each year is used for this purpose. According to one authority, "it has been proved from time immemorial that no bait is so attractive" (Wilcocks, 1883), and this may be due to its oily nature. Frampton (1954) says that long-

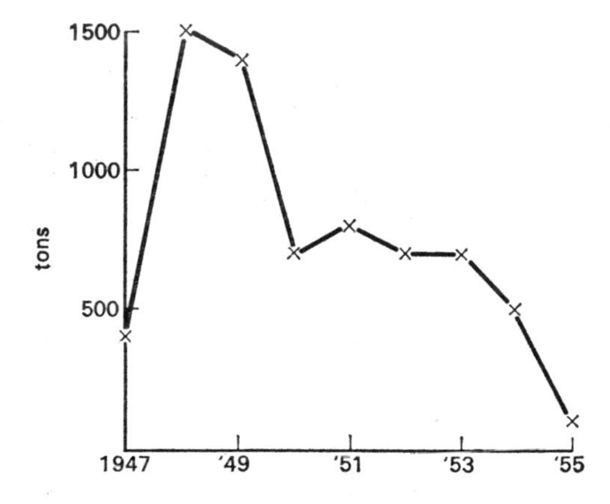

FIG. 11.1. Production and exports of salted pilchards, 1947 through 1955.

liners, fishing for skate, ray, plaice, dogfish, conger eels, and other genera, consider pilchards to be the best bait. Frequently they carry a few drift nets on board to catch pilchards before proceeding to the long-lining grounds. On occasion, if the pilchard fishermen do not use the fish for their own purposes, they can sell them at the quayside "by the hundred" to other fishermen for use as bait. The fish are generally cut into three before being placed on the hooks. The hooks themselves are at the end of short pieces of line about $1\frac{1}{2}$ fathoms long which are known as snoods. These in turn are attached to a main line of about half a mile in length and are placed along the main line at intervals of one or two fathoms.

Several of the long lines are laid, each anchored and marked with a dhan buoy. As about 7 miles of line are laid altogether, approximately fourteen of the main lines are set. The liners of the east coast lay more lines, up to 15 miles, and the snoods are further apart (Hardy, 1959).

Exports will now be dealt with in a little more detail. Export of canned pilchards does take place. The quantity, however, is very small, being in the region of 1 per cent of the total British production, and is not recorded statistically. The whole of the cured product is now exported, although as previously mentioned the amount is insignificant. The quantities exported from 1947 to 1955 are shown in Fig. 11.1. This shows the decline, referred to above, which occurred from 1951 onwards. The highest figure since the war was found in 1948, and was in the region of 1500 tons, being the highest export figure for this product since 1935. Before 1935, and back as far as 1918, all the figures were above this. Also, before the 1914 war the figures were generally above this level as far back as 1861. As can be seen in Fig. 11.2 the figures for exports of the cured product fluctuated violently from 1770, but averaged out at about 70,000 cwt (3560 tons). To compare this with the industry in recent times, if the total catch from the Channel is considered, this quantity has only been exceeded eight times since 1947. In the years prior to the First World War the seine net was used, and although this method gave violently fluctuating catches, the total catches tended to be considerable.

It was hoped at one time that any glut of pilchards which occurred could be transferred to a factory which would be able to produce animal feeding stuff. There is no factory situated

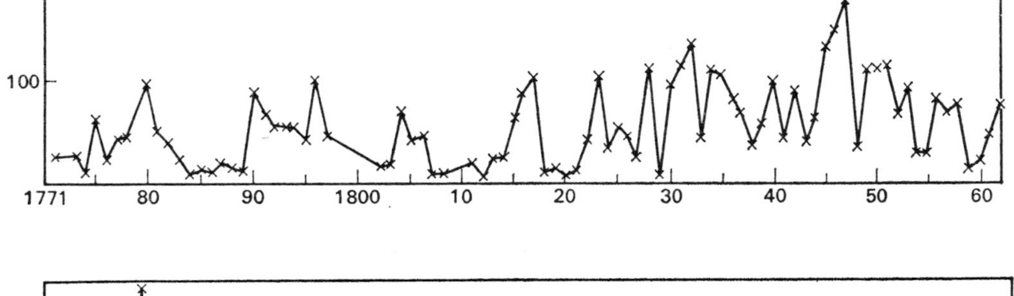

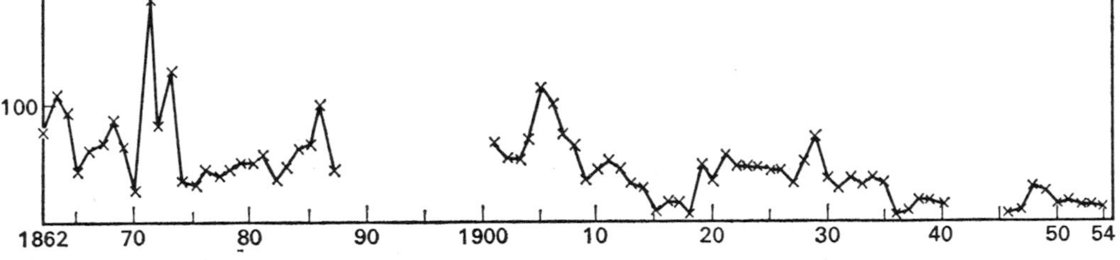

FIG. 11.2. Exports of cured pilchards from 1770.
(Figures in thousand cwt.)

in the West Country which produces fish meal. In 1954 the White Fish Authority were apparently thinking in terms of building such a factory in the hope that this would enable the fishermen to rely on a more regular and economic price for their catch. Whatever was done, or will be done, in this direction, nevertheless, cannot raise the price too far, as there is competition from South African and South West African imports, and the Cornish product could be priced out of the market. As the White Fish Authority stated in one of their reports (Great Britain, 1954), a fish meal factory in the south-west would "reduce the fishermen's fear of surpluses and encourage them to fish to the utmost capacity". However, the next report stated that the W.F.A. did not intend to "proceed with their project of erecting a fish meal and oil plant at Plymouth to serve the needs of the industry in the south-west of England". They were willing to assist a private company which was "endeavouring" to find the necessary equipment. Unfortunately, there is still no fish meal factory in the south-west "to serve the needs of the industry" in this way.

The fishermen's cry has always been that if they had the markets they could find the fish. Whether or not their claim is justified, and this point will be discussed later, it is clear that the markets available to them are too few and ought to be increased in number. This can only be done by increasing the number of ways in which the fresh product can be processed and by increasing the market for the fresh fish for its own sake.

CHAPTER 12

CONSUMER DEMAND

DEMAND for any product is influenced by a considerable number of factors. It is generally said that demand for a good is dependent on the size of the population, incomes, individual tastes and the background of the consumer, prices of other goods especially close substitutes, as well as the price of the good itself. When diagrams illustrating the demand of a product are drawn, the quantity demanded is generally plotted against price. The "demand curve" thus obtained is generally represented in Fig. 12.1 This illustrates the basic principle that, given normal conditions, it is usual for less of a product to be demanded, the greater the price.

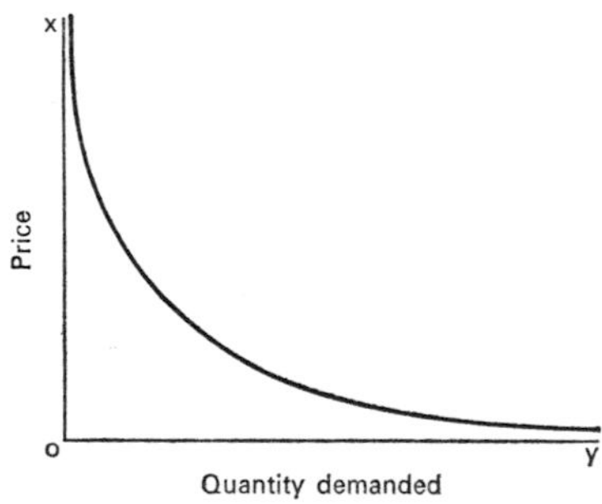

FIG. 12.1. Basic demand curve.

Demand for goods tends to fall into two basically different groups. Thus, it may be that the price of some goods can be altered over a wide range and yet produce only a slight change in demand. In this case demand is said to be inelastic. A commodity which falls into this group is generally vitally important to the consumer; bread may be quoted as an example. On the other hand, when a small change in the price of a commodity greatly alters the demand, demand for that good is said to be elastic.

In this chapter the demand for fish and some of the factors influencing it will be discussed. It should be mentioned here, however, that statistics and other data concerning this subject in general are very sparse and in connection with the pilchard industry are virtually non-existent.

Demand for fish cannot be stated categorically to be either elastic or inelastic, as for any one type of fish the demand will alter over a range of prices, being comparatively inelastic within certain price limits but becoming elastic at prices outside these limits. It is, nevertheless, possible to illustrate diagrammatically the way in which the demand for various types of fish may react to price changes. First, however, it was thought probable that fish in this country did not possess an homogeneous "public image", being perhaps unconsciously classified into several categories.

Although no definite divisions are made, there appears to be a tacit acknowledgement that two main classes of fish exist. These are:

1. fish bought quite frequently and regularly for use in the home, and
2. fish rarely bought for use in the home and which are associated with good class hotels and restaurants. Examples of fish found in this category are fresh salmon and trout, shellfish of various kinds, such as crabs, lobsters and oysters, and certain fish products such as caviar.

Cod, herring and various herring products, pilchard and pilchard products, would unhesitatingly be placed in the first part of the classification. However, between these and the second part of the classification (what might be termed, without dispute, the luxury fish) are certain species which may be considered by some to be a luxury, and by others to be used frequently in the home. The definition of "luxury" is subjective but even so it is possible to distinguish between "luxury fish" and "ordinary fish" and therefore the classification given above is still valid. The possible differences in demand–pattern will now be demonstrated.

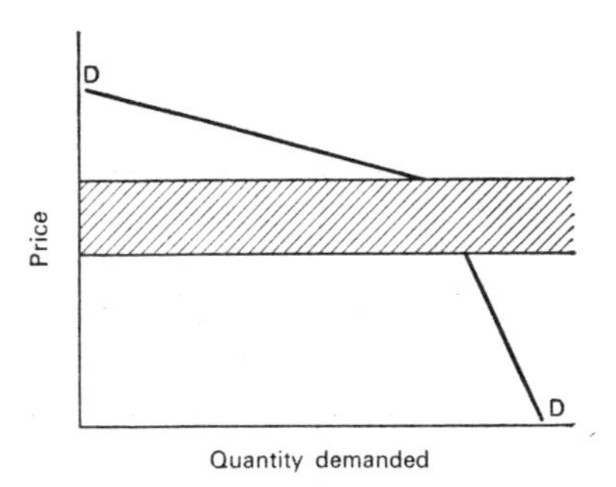

FIG. 12.2. Demand curve for "cheaper" fish.

For both classes of fish there is a certain area of the price range within which fluctuation can occur without greatly affecting demand. It is outside these limits, which will, of course, be relatively higher for the luxury goods, that the demand curves for the two types of fish are expected to differ.

(i) For the cheaper fish the demand curve is given in the graph, Fig. 12.2.

The shaded area represents the price range within which accepted fluctuations occur and in which there would probably be little alteration of demand with price changes. However, if the cost were to rise above these limits, the quantity bought would probably fall off rapidly, i.e. the proportionate change in demand is greater than the proportionate change in price which brought it about. Thus, in Fig. 12.2, above the shaded price range, demand is elastic. Below this price range the quantity bought rises as price falls. However, the proportionate change in demand is less than the proportionate change in price which caused it. That is, demand is inelastic.

(ii) In dealing with the luxury fish it is expected that the demand curve would appear as shown in Fig. 12.3.

Again, a range of prices within which fluctuations would not appreciably affect demand is shown shaded. When prices rise above this range the quantity bought would undoubtedly fall, but, because of the luxury nature of the product, the proportionate decrease in demand would be less than the proportionate increase in price. That is, even above the accepted price range demand is inelastic. Conversely, if the price were to fall somewhat below the accepted levels, then people who would not normally buy the luxury food would take advantage of the favourable situation and the consequent proportionate increase in demand would be greater than the proportionate fall in price, indicating an elastic demand for the product.

The demand for fish in general should be considered in relation to the demand for its substitutes as well, such as meat. Fish takes second place to meat in societies such as ours where the standard of living is reasonably high. Even with an increasing population the demand for meat is rising more than that for fish. In countries with a lower standard of living a different situation prevails. There is a need for a cheap nutritious food and this can generally be satisfied by fish, either fresh or processed in some way. As mentioned in Chapter 6, there used to be a considerable demand for salted pilchard in Italy. This product was ideal from the nutritional point of view as it contains about 22 g of protein for every 100 g

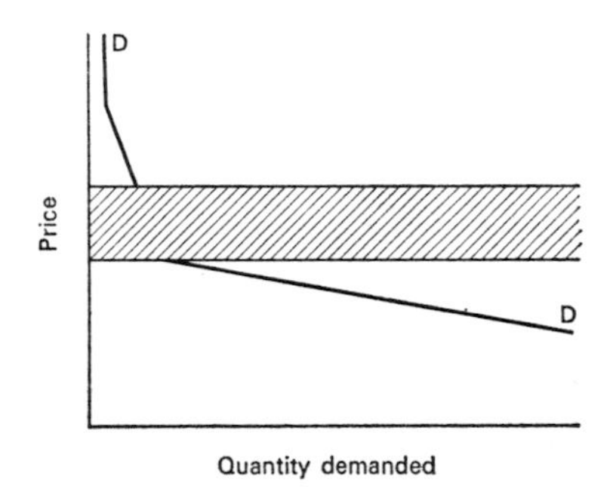

FIG. 12.3. Demand curve for "luxury" fish.

eaten, and this compares favourably with beef which has about 21·3 g protein for every 100 g. Furthermore, of the weight purchased, 100 per cent of the pilchard product would be edible, but even considering beef sirloin without bone, usually only 60 per cent of the weight purchased is edible. The calorific value of pilchards is about 30 per cent less than that of beef, but the price of the latter is very much higher (Baker and Foskett, 1958). From the point of view of the pilchard industry, it is an unfortunate result of an improved standard of living that the demand for fish in general, and of the cheaper products derived from them in particular, usually falls.

In Great Britain a survey (Great Britain, 1961a) concerning the feelings of the average consumer towards fish was carried out in the early summer of 1960. The results showed that fish was served in practically every household and eaten by the vast majority of the population. As the report states, there therefore appears to be "little prospect of gaining new fish eaters". Nevertheless, there should be plenty of opportunity to encourage those who eat fish to eat it more often, particularly those in the lower income and age groups by whom it is least eaten. At first sight it seems anomalous that the survey should find that the people in the lower income and age groups should be those who eat least fish. However, this is probably a legacy from the economic conditions of the late 1920s and 1930s when it was the people in the lower income groups who were forced to eat fish, because the price of meat was prohibitive. Now they can afford meat they eat it in preference to fish, and their children have, therefore, not acquired the fish-eating habit. It should be possible to persuade the younger age groups to eat more fish by better advertising and "education" in the methods of its preparation. This would enable them to make a meal of fish more appetizing. This aspect is of extreme importance to the future of the fish industry in Britain, as another of the findings of the report was that there is a general impression that fish is "a dull and uninteresting food" which engendered a feeling of indifference towards it. Whilst dealing with the consumer survey, it should be noted that the most serious criticisms levelled against fish referred to the fluctuations in its price and uncertainties about its freshness.

Another consumer survey (Taylor, 1960) showed how important the factors of individual tastes and background were on the demand for fish, and how these differed from region to region even within England. In the northern part of England, east of the Pennines and in the

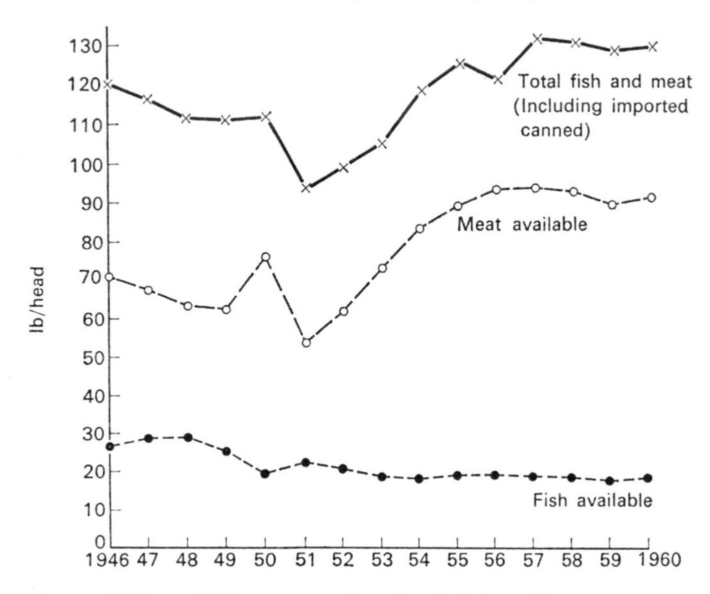

FIG. 12.4. Meat and fish available per pound per head of the British population, 1946 through 1960.

East and West Ridings of Yorkshire, for example, approximately 2·9 oz of fresh white fish are eaten per head per week. This compares with only 2·5 oz per head per week in south east and southern England (excluding London). For cooked fish the figures are 1·42 oz and 0·63 oz respectively. The first set of figures given are not necessarily very significant as in the northern region there is probably a greater supply of fresh fish than in the southern region. However, the second two figures show a definite difference in consumer preferences between the two regions.†

When the amount of fish available per head of the British population per annum is compared with the amount of meat available per head of the population, from 1946 to 1954 it is found that the two move inversely, except for the year 1949 (Fig. 12.4). From 1946 to 1952, the amount of meat available (except for 1950) can be seen from the graph to be on a generally lower level than it has been since 1953. Conversely, the amount of fish available from 1946 to 1951 (again with the exception of 1950) was on a higher level than it had been since 1953. The figures shown in Table 15 give the quantities (in pounds) of fish

† These regions and others in the country are fully defined by Taylor.

and meat available per head of the British population per annum for the two periods, 1946–52 inclusive, and 1953–60 inclusive (Great Britain, 1957–61).

TABLE 15. COMPARISON OF QUANTITIES OF MEAT AND FISH AVAILABLE PER HEAD FOR TWO PERIODS BETWEEN 1946 AND 1960

	1946–52	1953–60
lb meat per head p.a. available	65·08	88·4
lb fish per head p.a. available	24·4	18·6

It seems, therefore, that when the amount of meat available was relatively low, changes in the quantity available per head of the population were rapidly reflected by inverse changes in the amount of fish available. Since meat has been readily available, the quantity of fish available has been governed, at least in part, by demand for fish and meat combined, and the movements of the two have thus tended to be correlated. The situation is more complex than represented here, and a considerable number of factors is concerned with causing the pattern of supply.

Amongst these other factors, the availability of the biological resource is of primary importance. If the pilchard fishermen, as they claim, can always get the fish if there are markets open to them (that is, they consider the biological resource always easily available), why did the number of canning factories taking pilchards drop from nine to three in 10 years? The demand for the raw material had at one time been there, but had obviously not been fulfilled. (It was not due to a fall in demand for the canned product that the demand from the factories declined as can be seen by the level of imports of canned pilchards.)

Seasonal Changes in Demand

Judging by the number of ounces of fish eaten per head of the population per week, averaged in each quarter from 1956 to 1960, there is no unswerving pattern of seasonal changes in demand. But from the graph (in Fig. 12.5) it appears that there is usually a fall in the quantity of fish eaten during the winter. (That is, in the first quarter and sometimes the last quarter of each year.) Where this does not occur, in the last quarter of 1958 and the first quarter of 1959, it is found that the amount of fish available in 1958 and 1959, although falling slightly, did not in fact fall as much as the amount of meat available. (See graph giving the total fish and meat available per head of the population per year: Fig. 12.4.) Furthermore, as shown in Table 16, the quantity of meat consumed per head of the population was not as great in the first quarter of 1959, when compared with the corresponding quarters of 1956, 1957, 1958 and 1960 (Great Britain, 1957–61).
So here again it seems that a smaller consumption of meat per head had boosted the consumption of fish.

TABLE 16. MEAT CONSUMPTION IN OUNCES PER HEAD OF POPULATION

First quarter of:				
1956	1957	1958	1959	1960
19·08	19·44	19·24	17·83	17·97

When considering consumption of particular types of fish, there are no figures dealing specifically with pilchards. However, if those for a similar type of fish, the herring, are considered, there is a regular pattern of the amount consumed during each quarter of the year. It can be seen from the graph (see Fig. 12.5) that the least amount of herring is consumed during the second quarter of the year, and that during the third and fourth quarters the quantity increases somewhat. There is again a slight decrease in the first quarter of each year (compared with the last quarter of the previous year). It can also be seen from the graph that the pattern of herring consumption moves independently of the fresh fish consumption pattern. It is likely that the pattern for the herring is largely caused by the landings of herrings down the east coast, although in recent years the contribution to the total catch by west coast fisheries (especially the North Minch) has also been important (Great Britain,

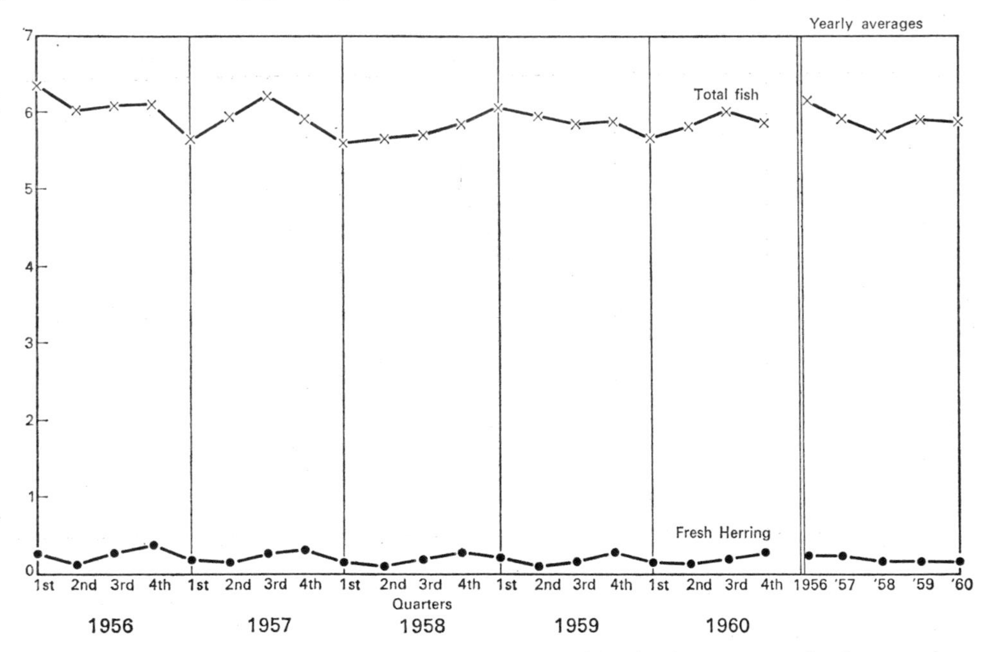

FIG. 12.5. Average quarterly consumption of all fish and herring in ounces per head per week, 1956 through 1960.

1960a, 1961b). Thus it is not until May and June that herrings in the North Sea start increasing in quality by feeding on plankton, and this improvement continues into July and August. In August, also there is an influx of spawning fish to the north-east coast of Scotland. From August to the end of the year shoaling fish are found progressively southwards down the east coast, and during the last quarter of the year the majority of herring drifters from England and Scotland are based at Yarmouth and Lowestoft (Hodgson, 1957), thus bringing the amount of herring available to its maximum.

It is most probable, although as already stated no figures to support the theory are available, that the consumption of fresh pilchards in the West Country is similarly seasonal and based on the catching-pattern outlined previously. Thus, it would be expected that the smallest quantity of pilchards would be eaten during the first and last quarters, and the maximum reached in the months of the third quarter of the year.

However, the majority of pilchards eaten in England are canned, so it would appear to be more pertinent to study the seasonal consumption pattern of canned fish during the year. When consulting the statistics available, it is found that the breakdown, as would be expected, is not very detailed. There are, however, figures for consumption of canned and bottled fish, in ounces per head per week for each quarter of the year, until 1960. In 1960 the statistics alter. No figures for bottled fish are mentioned and those for canned fish are broken down into figures for canned salmon and "other" canned fish (Great Britain, 1957–61). In the graph (Fig. 12.6) both the figures for the "other" canned fish (bottom line) and the total of this and canned salmon (top line) are given for 1960.

In 1956 and 1957 and again in 1960 there is a distinct pattern in the graph showing an increase in consumption of these foodstuffs in the second quarter of the year, followed by a decrease in the last two quarters of the year. (Although in 1957, the amount consumed in the second and third quarters of the year was the same.) In 1958 the first three quarters of

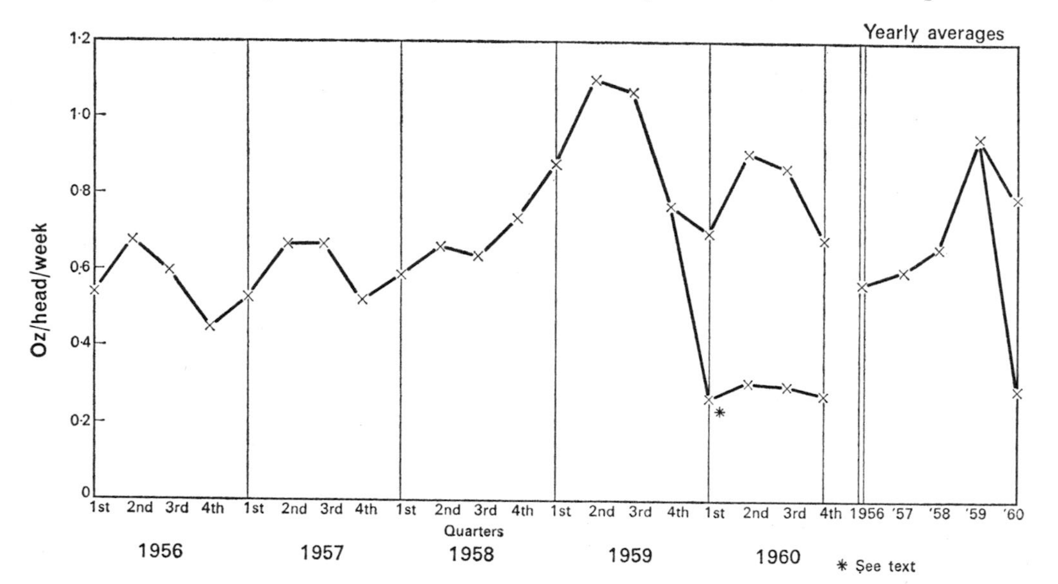

FIG. 12.6. Quarterly consumption of canned and bottled fish in Great Britain, 1956 through 1960.

the year follow the pattern of the previous 2 years, but in the last quarter, instead of the decrease observed in the previous 2 years, there is a large increase. This increase is continued into the first and second quarters of 1959 (where the maximum of 1·01 oz per head per week is reached) until in the third and fourth quarters there is the more usual reversion to a decline. In 1960 the pattern for the consumption of the total amount of canned fish, including salmon, and for the other canned fish both follow what can probably be regarded as the normal seasonal pattern.

The large increase of consumption of canned fish, noted in the last quarter of 1958 and through into 1959, was probably due to the removal of the quota restrictions on canned salmon in 1958. The annual average of the consumption of canned fish is also shown on the graph (Fig. 12.6). Apart from the very large increase in 1959 (already explained above), there has been a more or less steady increase in consumption of canned fish from 1956 onwards.

An interesting point to note here is that the seasonal patterns for the consumption of fresh herrings and the consumption of canned and bottled fish are inversely correlated. The significance of this possible interchange, of fresh herring with canned fish of some sort, needs to be determined by further detailed consumer study. However, probably what is significant is that from a comparison of the figure showing the consumption of fresh fish (see Fig. 12.5) with that showing the consumption of canned fish (Fig. 12.6) it appears that canned fish are not eaten as an alternative to fresh fish. If the two were considered as alternatives in the mind of the housewife the two consumer-pattern graphs would fluctuate inversely with respect to one another. As this is not the case, canned fish must hold a firm place of its own in the diet of most households.

One final interesting tendency may be observed in these graphs, and it may be that the pilchard industry will benefit from this tendency in the future. This is, that although the average yearly consumption of fresh fish, both white and herring, has been undergoing a gradual decline (from 1956 to 1960 inclusive), the annual average quantity of canned fish eaten during these years has shown a steady increase. As mentioned earlier, there was a decline in 1960 after the large increase in 1959, thought to be due to a rush on canned salmon. Nevertheless, considering the average for 1959 to be unnaturally elevated, it certainly seems that canned fish is becoming eaten more frequently in this country, although to what extent this applied to canned pilchards cannot be fully ascertained from the statistics available. It does seem that consumer demand of the pilchard, judging by the figures of imports of canned pilchards (shown in Fig. 7.2) has stayed at a comparatively high level in recent years. The increase in demand for canned fish is associated with a greater consumption of what are known as convenience foods, throughout all income-groups. These convenience foods include, as well as canned and bottled fish, canned and ready cooked meats, canned fruits and soups, cakes and prepared puddings, ice-cream when served at a meal and breakfast cereals.

According to the Annual Report of the National Food Survey Committee for 1960 (Great Britain, 1960b), expenditure on convenience foods in all income groups has increased since 1956, but this has been most marked in the highest and lowest income groups. As the middle income groups have traditionally been the largest spenders on convenience foods, it can be seen that differential demand for these products between the classes has evened out. Therefore, as the reports state, "reliance on convenience foods is becoming less associated with social class".

The report also states, quoting another source (Crawford and Broadley, 1938) that it was mostly "the lower-middle and the more prosperous working-class homes" which bought canned salmon before the war. However, in 1960 the greatest demand for this product came from the highest income groups, and this was only slightly greater than the demand for this product by other income groups. This is another example that more people representing all income groups are eating canned fish.

In conclusion it may be said that for most types of fish demand increases either when meat is not so readily available, or when it is too expensive. There is a basic level of demand for fish but this has had a tendency to decrease gradually during the past few years. The fluctuations in price which occur, unless very violent, do not appear to be a major factor in bringing about alterations in demand. This stability which has been noted is probably brought about by a habit of eating fish at certain times (for example, on Fridays) and it is likely that the frequency with which fresh fish is eaten could be increased by "education" and advertising, although, for the pilchard, completely new channels of marketing and distribution would have to be opened up first.

CHAPTER 13

TRENDS IN SOME WEST COAST PORTS

In 1962 there were eight ports in the West Country of England which had pilchard fishing boats operating from them. Of these, seven were in Cornwall, Plymouth being the only port connected with the industry in Devon. Newlyn, Looe and Mevagissey were easily the three most important in 1962, but in the short time since then the picture has changed considerably and the following pages will indicate the decline that has occurred throughout the industry and will show that it is not now possible to say with certainty which are the major ports of the industry. The catches and prices of pilchards on first landing in Newlyn, Looe and Mevagissey will also be shown, and finally the most recent changes and trends in these ports will be discussed.

The numbers of pilchard drifters at seven Cornish ports (and at the Devonshire port of Plymouth), and the numbers in each length class have been tabulated for 1962 and 1966 (Tables 17a and 17b) and the data have also been illustrated cartographically in Figs. 13.1 and 13.2.

It can be seen that without exception the number of pilchard vessels in all the ports has declined rapidly in 4 years and those most noticeably affected are the ports which were the most important in 1962. Furthermore, Plymouth, Portloe and St. Ives now have no boats fishing for pilchards. The overall decline in numbers of vessels operating out of Cornish ports for pilchards has been from a total of seventy-five in 1962 to a total of twenty in 1966. This decline in itself is very rapid but the overall decline since the mid-1950s, when the industry was at its most flourishing in the post-war era, is even greater. It was estimated that in 1954 at the height of the season 150 boats were fishing for pilchards (Frampton, 1954).

Also indicated in Tables 17a and 17b and in Figs. 13.1 and 13.2 is the relative importance of pilchard fishing to the various ports concerned. For example, in 1962 the thirteen pilchard

Table 17a. To Show Numbers of Pilchard Drifters of Various Length Classes at the Ports, and Expressed as a Percentage of all Other Boats at the Ports to Indicate Relative Importance in 1962

Port	20′	20–29′	30–39′	40′	Total	% of all boats
Plymouth	–	2	3	–	5	22
Looe	–	2	6	5	13	61
Polperro	–	–	4	–	4	50
Mevagissey	–	9	4	4	17	58
Portloe	3	–	–	–	3	43
Newlyn	–	3	4	11	18	25
St. Ives	–	–	–	7	7	29
Porthleven	–	1	3	4	8	47

(Data of numbers of boats in each class obtained from Great Britain, 1962a.)

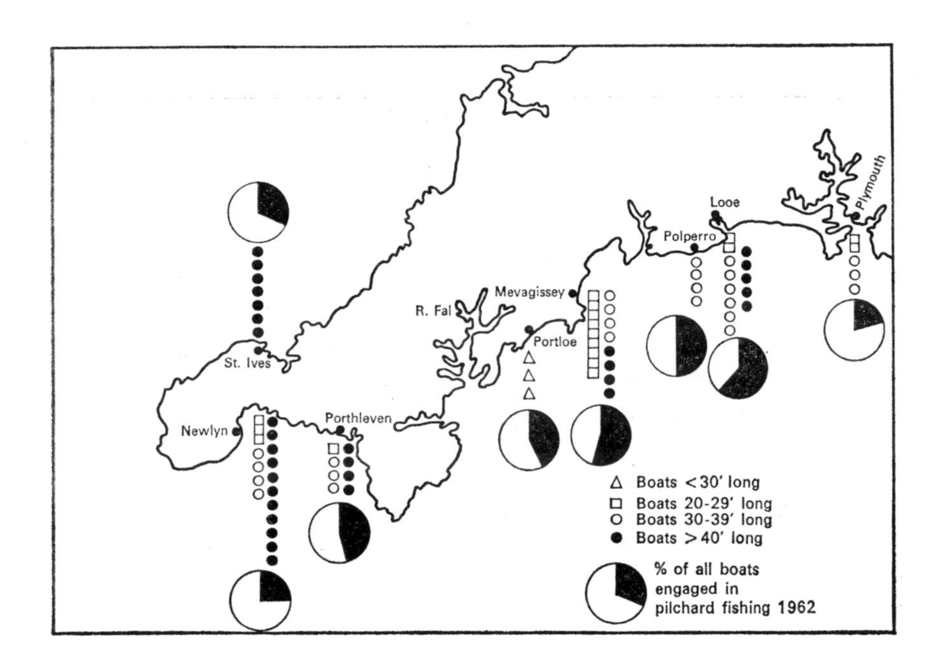

FIG. 13.1. Number of pilchard drifters at various English ports in 1962.

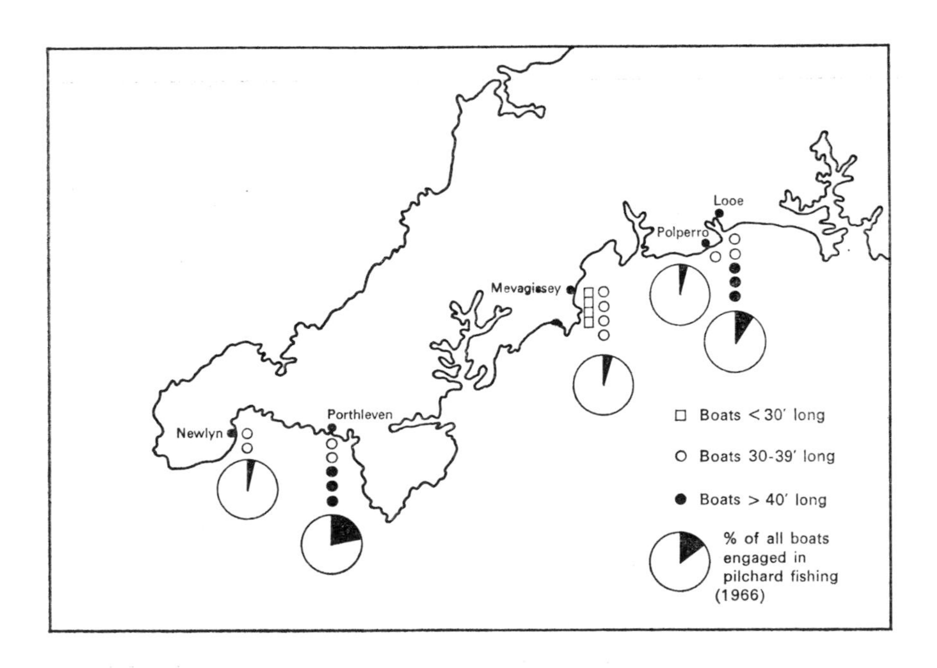

FIG. 13.2. Number of pilchard drifters at various English ports in 1966.

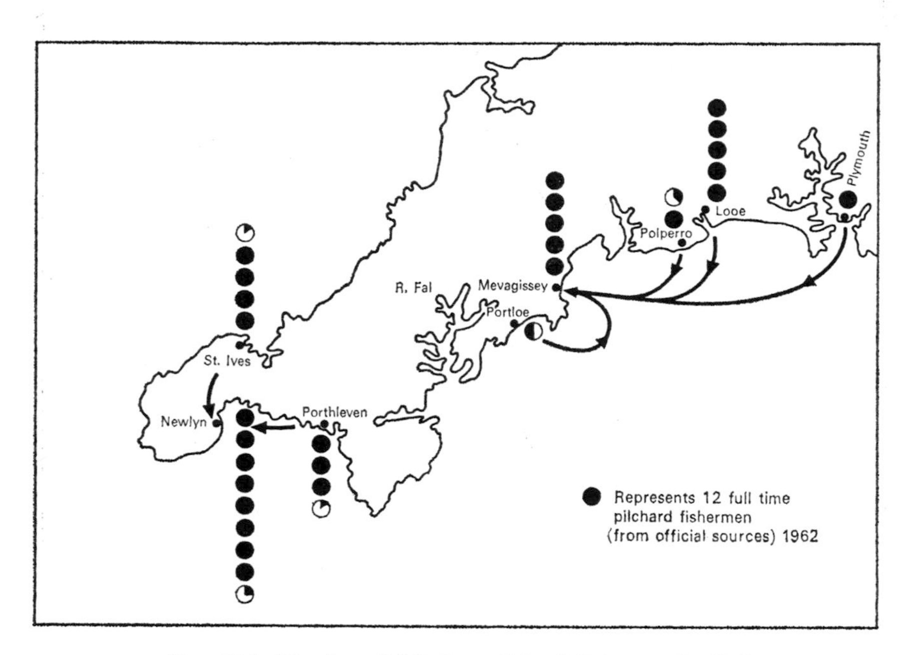

FIG. 13.3. Number of full-time pilchard fishermen in 1962.

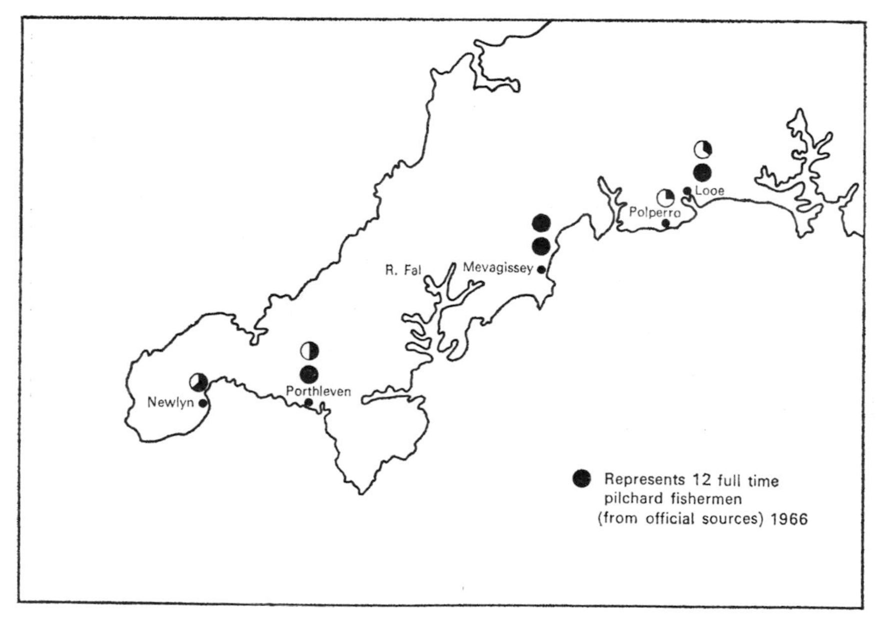

FIG. 13.4. Number of full-time pilchard fishermen in 1966.

TABLE 17b. TO SHOW NUMBERS OF PILCHARD DRIFTERS OF VARIOUS LENGTH CLASSES AT THE PORTS, AND EXPRESSED AS A PERCENTAGE OF ALL OTHER BOATS AT THE PORTS TO INDICATE RELATIVE IMPORTANCE IN 1966

Port	20′	20–29′	30–39′	40′	Total	% of all boats
Plymouth	–	–	–	–	–	—
Looe	–	–	2	3	5	11
Polperro	–	–	1	–	1	8
Mevagissey	–	4	3	–	7	8
Portloe	–	–	–	–	–	—
Newlyn	–	–	2	–	2	2
St. Ives	–	–	–	–	–	—
Porthleven	–	–	2	3	5	21

(Data of numbers of boats in each class obtained from Great Britain, 1966a.)

boats at Looe represented a slightly greater proportion of the total number of boats at that port (61 per cent) than the seventeen at Mevagissey (58 per cent). A similar relative importance is suggested by the 1966 figures which give the five Looe pilchard boats as 11 per cent of the total number of boats at the port and seven Mevagissey pilchard boats as 8 per cent of the total. At Newlyn, the major fishing port of the area, although there were eighteen pilchard boats based there in 1962, these represented only 25 per cent of the fishing boats operating from the port. By 1966, however, there were only two pilchard drifters at Newlyn—2 per cent of the total number of vessels based at the port.

Correlated with this decline in the number of pilchard drifters is the decline which has occurred in the number of people finding employment connected with the industry. In 1954 there were about 500 men engaged in pilchard drifting (Frampton, 1954), by 1962 this number had fallen to 334 working on pilchard boats full-time. In 1966 sixty-nine were employed pilchard drifting and none of these could be considered as fully employed on pilchard boats. Of the numbers of workers finding employment in the associated industries of canning and curing 400, some of whom would be seasonal labour, were estimated to be employed at the height of the season in 1954. In 1962 and 1966 the totals had dropped to 110 and 98. There does not seem to have been a proportionate decrease in the labour force here, but the figures for the total number of people employed include part-time workers and, in fact, conceal a rather greater decrease. On breaking the figures down into full-time and seasonal employment a decline from seventy full-time workers in 1962 to forty-four in 1966 is revealed, and the number of part-time staff rose from forty in 1962 to fifty-four in 1966.

Figures 13.3 and 13.4 show the number of full-time pilchard fishermen at the eight ports in 1962 and 1966 respectively. In some cases, the number of men at a port may appear not to be correlated with the number of boats located there. This is because the number of men is dependent on the length of the boat as well as the number of boats; small boats no more than 20 ft in length need only a two-man crew, whereas boats of 30 ft and above need at least three and usually four men in the crew.

Also shown in Fig. 13.3 are the centres to which fish would normally be sent for processing. It can be seen that in 1962 there were only two ports at which processing of pilchards occurred. These were Mevagissey and Newlyn, the cannery at Looe having stopped taking pilchards earlier in 1962. Until recently Mevagissey and Newlyn had both canning and curing facilities, but in 1965 the curing plant at Mevagissey closed down so that now (1967) there is one canning factory at Mevagissey and there are two canning firms and two small curing plants at Newlyn. If the processors at either of these two ports are unable to take any more fish, then Henry Sutton's factory at Great Yarmouth will probably take them. It sometimes happens that a glut of pilchards occurs so that some of the catch cannot be taken by any processors. When such occasions arise the fish have to be dumped in the sea because, as mentioned in Chapter 11, no fish reduction plant exists in the area.

If the total yearly catches for Newlyn, Looe and Mevagissey are briefly considered from the beginning of the century some interesting points emerge. If one takes a sample of the catches for these three ports for example for 1908, 1918, 1928, 1938, 1948 and then for 2-yearly intervals through 1964. It can be shown that for most of the sample years (eight out of thirteen) the catch of pilchards landed at Newlyn has been greater than the catch landed at the other two ports. However, since the last war, Looe and Mevagissey have gradually been increasing in importance with reference to the pilchard industry, but on the whole this does not appear to have been at the expense of Newlyn. Catches at the other, smaller ports have fallen with the tendency towards increasing centralization and concentration of fishing effort noticed earlier (Chapter 6). When comparing landings at Looe and Mevagissey it appears that the former was the more important, at least until 1956. During 1958, 1960 1962 and 1964, however, a greater quantity of pilchards has been landed at Mevagissey than at Looe.

The tendency for more fish to be landed at Mevagissey than at Looe in recent years may be connected with the difference in natural facilities of the two ports. Looe is situated at the mouth of the river Looe which is only navigable for vessels of fishing size for the few hours around high water. This means that fishing trips may be unnecessarily prolonged. Furthermore, even although there are 1800 ft of quayside there, the depth at the quay is only 13 ft at high water on an average spring tide, and is virtually dry at low water. Mevagissey, on the other hand, has an inner and an outer harbour. The inner harbour, although it dries out at low tide and is only one-third of the area of the outer harbour, is capable of accommodating all the boats at the port, and the outer harbour, facing east and having granite piers on its north and south sides, affords shelter from most winds. Even at low water, the level at the pier does not fall below 10 ft on an ordinary spring tide (Kennea, 1957). Fishing boats can, therefore, return to Mevagissey as soon as their mission is completed.

The better natural facilities of Mevagissey have existed since the building of the outer harbour there in 1865. Now, however, Mevagissey also offers better facilities for the processing of the fish. The Looe curers have not been operating since 1955 and the canning factory stopped accepting pilchards in 1962. Thus it is probable that landings increased at Mevagissey in 1956 because of the closure of the Looe curers; and now the landings at Mevagissey should be greater than before due to the stimulus provided by the closing down of the Looe canning plant. At Mevagissey the salting factory probably salted down more pilchards during the 1950's than that at Newlyn (Great Britain, 1962), and this, coupled with the fact that the canning factory is still in operation there makes it a port of outstanding importance to the pilchard industry second only to Newlyn.

The recent closing of the curing plant at Mevagissey may alter the interrelationships of these three ports once again. In the light of the size of the landings at Newlyn, Looe and Mevagissey, it is difficult to explain why the canning plant at Looe, and the curing plant at Mevagissey have both had to shut down since 1962. The landings have remained fairly steady at both these two ports and yet at Newlyn, where there are two canning and two curing plants, the landings have fallen consistently and sharply since 1962. The landings, with their values, for the three ports for 1962 through 1966 are shown in Table 18.

TABLE 18. LANDINGS AND VALUE OF PILCHARDS, 1962 THROUGH 1966, AT NEWLYN, LOOE AND MEVAGISSEY (Great Britain, 1966a)

Port	Year	Landings (cwt)	Value (£)
Newlyn	1962	16,161	21,137
	1963	9,986	12,113
	1964	8,423	10,235
	1965	3,758	4,852
	1966	2,872	4,492
Looe	1962	8,133	11,403
	1963	12,600	15,116
	1964	8,609	11,183
	1965	10,144	11,222
	1966	14,508	14,693
Mevagissey	1962	13,022	15,624
	1963	14,158	17,772
	1964	13,593	16,337
	1965	9,746	11,723
	1966	7,902	10,171

One possible explanation is that there has been a more consistent demand for pilchards from the processors at Newlyn than elsewhere so that fish landed at other ports have been transported to the Newlyn plants. The fall in landings of pilchards at Newlyn itself is probably a natural result of a declining industry whereby skippers, reaching retiring age, have left the industry and have not been replaced by other, younger men; the number of pilchard boats there in 1962 was eighteen and in 1966 only two. Other skippers have probably changed to trawling which has been quite flourishing in the Western Channel since 1963.

Seasonal Fishing Pattern

As seen in Chapter 7, the total pilchard catch in recent times has not usually exceeded the catch in pre-war years, although it has tended to be more regular, that is less seasonal in nature.

At one time no fishing for pilchards occurred during the first 6 months of the year, and it was very rare in the last 2 months of the year. This meant that the whole of the fishing season was concentrated usually into 4, but sometimes into 5 months. For example, in 1908 the total amount of pilchards landed at Newlyn (1563 tons) was all caught within the 4 months from July to October. Fishing at Looe and Mevagissey continued into November of that year but the quantities landed in those 2 months were insignificant (4 tons and 2 tons respectively). Although, at the time, fishing for pilchards by seining was still being carried on, probably at all three ports, it is certain that the number of drifters would exceed the number of seines at all three ports. So this late summer and early autumn fishery was characteristic of the early drifting fishery: a pattern of events inherited from the seiners. In 1918, probably due to the effects of the war and to the fact that seining would no longer be taking place, the catches at these three ports were very low. However, at Newlyn the fishery was again concentrated into the months from July to October, whilst at Looe and Mevagissey all the pilchards were landed in October and November.

Since the late 1920s there has been a tendency to have a longer fishing season, starting earlier in the year, but this has not been at the same time, nor to the same extent at each port when judged by the landings at the three major ports of Newlyn, Looe and Mevagissey. Neither can fishing be said to have started early at Newlyn, followed by Mevagissey and then by Looe as the shoals moved round the coast. Thus, in 1948, for example, landings at Mevagissey during the first 4 months of the year were less than half those at Looe (59 tons and 136 tons respectively) and those at Newlyn were negligible, amounting to 18 tons. Since about 1950, however, there has been a tendency at both Looe and Mevagissey for landings of pilchards to occur throughout the year. February, March, April, October, November and occasionally December have in different years and at different ports given very small landings or none at all. Fishing during those 6 months mentioned is very greatly affected by weather conditions as the figures in Table 19 show. They are the quantities (in stones) of pilchards bought by a canning firm during the months November to February

TABLE 19. INDICATING THE EFFECT OF WEATHER ON WINTER LANDINGS AT NEWLYN

(quantities in stones)

Year	November	December	January	February
1957–58	N.A.	6,299	35,080	25,785
1958–59	3,997	5,841	32,017	11,897
1959–60	33	Nil	24,321	11,753
1961–62	156	18	750	1,759
1962–63	1,213[a]	214	Nil	N.A.

[a] Includes one catch of 409 stones.
N.A., figures not available.

inclusive for the 3 years from 1957–8 to 1959–60 compared with the two years 1961–2 and 1962–3 when weather was very difficult and exceptionally cold.

One other interesting fact emerges from a study of the monthly landings at these three ports for the past 10 years. At Newlyn, far more than at Looe and Mevagissey, pilchard drifting has remained a seasonal occupation. Landings of pilchards at this port have been non-existent or negligible during the first 3 or 4 months of the year at Newlyn and frequently again during November and December. To give recent examples the figures for 1961 and 1965 are shown in Tables 20a and 20b.

There is no absolutely definite reason for this. Probably the most powerful causative agent of such a pattern is the tradition of the industry. For hundreds of years pilchard fishing has been a seasonal occupation and, therefore, tends to remain so. Beyond this, however, a greater variety of fish is caught by the vessels fishing from Newlyn so that the seasonal nature is probably somewhat exaggerated. The pilchard vessels at Newlyn at least in former times, were larger generally than at other Cornish ports (see Table 17a) and are thus more

TABLE 20a. SHOWING MARKED SEASONALITY OF PILCHARD LANDINGS AT NEWLYN WHEN COMPARED WITH LOOE AND MEVAGISSEY, 1961 (Great Britain, 1962a) (quantities in tons)

1961	Newlyn	Looe	Mevagissey
January to April	4	270	485
May to September	1021	227	338
October to December	1·6	13	20

TABLE 20b. SHOWING MARKED SEASONALITY OF PILCHARD LANDINGS AT NEWLYN WHEN COMPARED WITH LOOE AND MEVAGISSEY, 1966 (Great Britain, 1966a) (quantities in tons)

1966	Newlyn	Looe	Mevagissey
January to April	14	78	20
May to September	126	500	94
October to December	5	10	284

versatile. This enables the fishermen to go out for other kinds of fish when the quality of pilchards is not at its best, and when the other fish, such as skate and rays, are fetching good prices. Thus, in the first 4 months of 1961, for example, considerable quantities of skates and rays were landed at Newlyn and these fetched £4 10*s*. per hundredweight. The small quantity of pilchards landed during the same period fetched only £1 7*s*. per hundredweight. The exact figures are given in Table 21.

TABLE 21. COMPARISON OF CATCH AND VALUE OF PILCHARDS WITH SKATE AND RAYS AT NEWLYN FOR JANUARY THROUGH APRIL, 1961 (Great Britain, 1962a)

	Quantity landed January through April	Value	Price/cwt
Pilchards	84 cwt	£112	£1 7*s*.
Skates and rays	3229 cwt	£15,635	£4 10*s*.

The fishermen who are able to turn over to long-lining are, therefore, going to do so, and they will not fish for pilchards when such good prices are being fetched by the other species.

Prices

The selling of the pilchard differs from that of most other fish in that it is not auctioned at the quay-side or at the market. The usual procedure is for the canners to make a contract with the fishermen for a certain (or unspecified) amount of fish at a pre-agreed price per stone. Figure 13.5 shows the changes in price that have been obtained by the fishermen for pilchards on first landing from 1951 through 1966. From 1951 through 1956 the prices fluctuated quite widely, reaching high peaks in 1952 and 1956 of 3*s.* 10*d.* and 4*s.* 0*d.* per stone respectively. From 1957 through 1961 the prices fluctuated, there being three troughs and two peaks and each successive trough and peak being lower than the previous one. In 1962 there was another slight rise in price, but generally there has been an overall downward trend since 4*s.* 0*d.* was reached in 1956 to the 2*s.* 6*d.* per stone in 1966. There is an interesting correlation here between the variations in the price paid to the Cornish fishermen and the quantity of pilchards imported into the country. The patterns of the two factors from 1957 through 1965 are identical.

Thus, in the years when more pilchards are brought into the country the price obtained by Cornish fishermen increases. Therefore, either the canners are prepared to pay more when an increase in demand is indicated by increasing imports, or there is a genuine increase in demand during those years when both factors rise together.

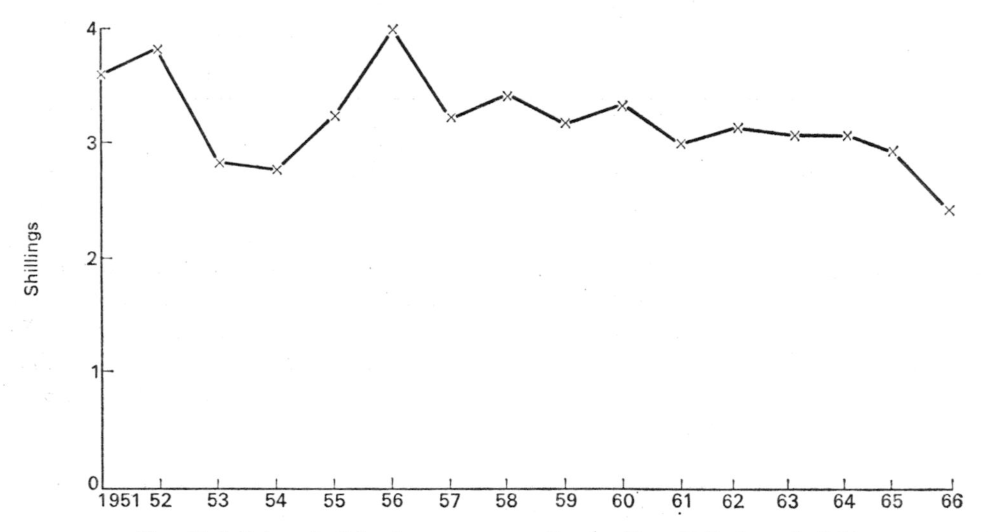

FIG. 13.5. Price of pilchards per stone on first landing; 1951 through 1966.

Whatever the true explanation might be, and this cannot be ascertained from the statistics at present available, it appears that the Cornish pilchard industry need not fear that the South African imports have the effect of lowering their prices. In fact the opposite view can be put forward, as from a study of the statistics it could be argued that the imports are actually helping to raise the price of the home product. Nor do the statistics show a slump in the Cornish industry with rising imports, as shown in the graph giving both imports and home production (Fig. 7.2).

Prices at the various ports, although changing from month to month (and in fact from port to port) have been showing a tendency recently to remain constant from month to month in any given port. For example, in 1956 the price paid to the fishermen for pilchards landed at Mevagissey was 3*s.* 6*d.* per stone from April to December (Fig. 13.7), and in 1961, whenever pilchards were landed at Looe they fetched 3*s.* 2*d.* per stone (Fig. 13.8). The graphs in Figs. 13.6–13.8 also suggest that there has been a tendency during recent years for the differences in the price per stone of pilchards paid at the three major ports to become less marked. Newlyn prices also seem to have become less independent of the general price trend (except for the sudden rise in price per stone in December 1956).

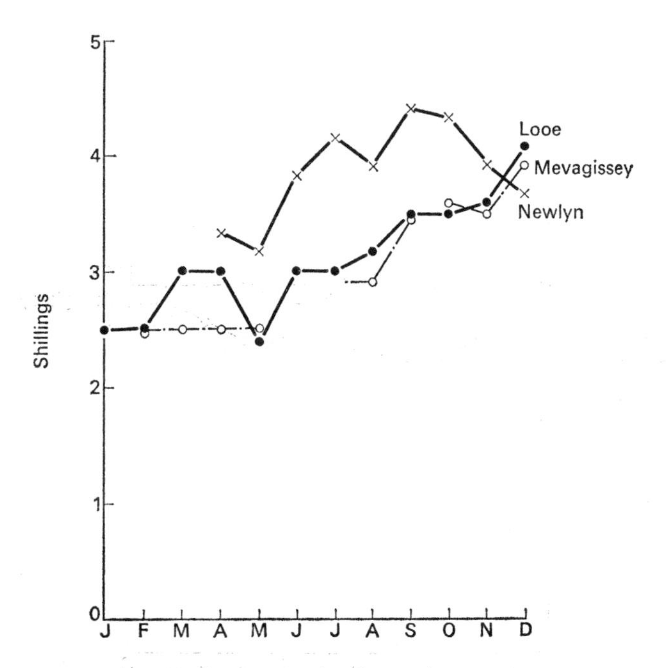

FIG. 13.6. Monthly prices of pilchards per stone at Newlyn, Looe and Mevagissey, 1951.

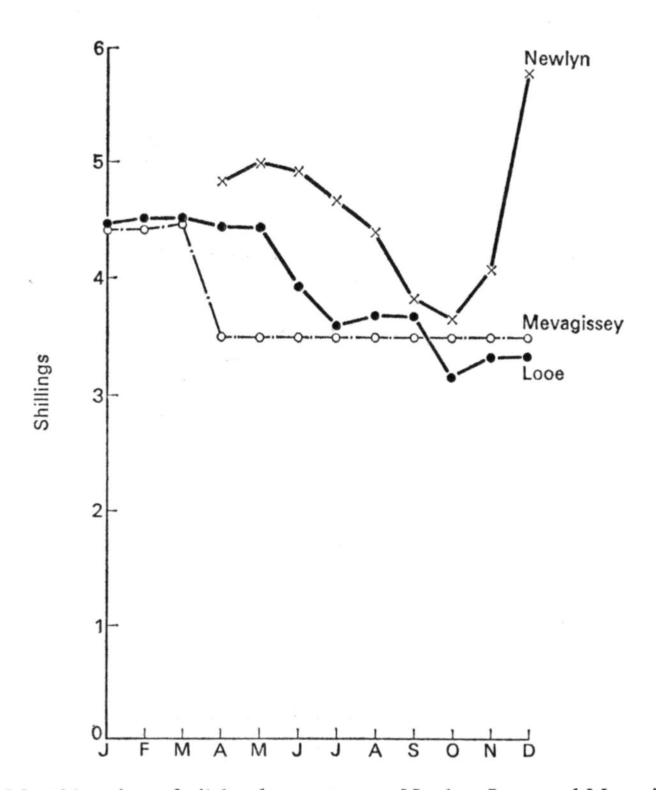

FIG. 13.7. Monthly prices of pilchards per stone at Newlyn, Looe and Mevagissey, 1956.

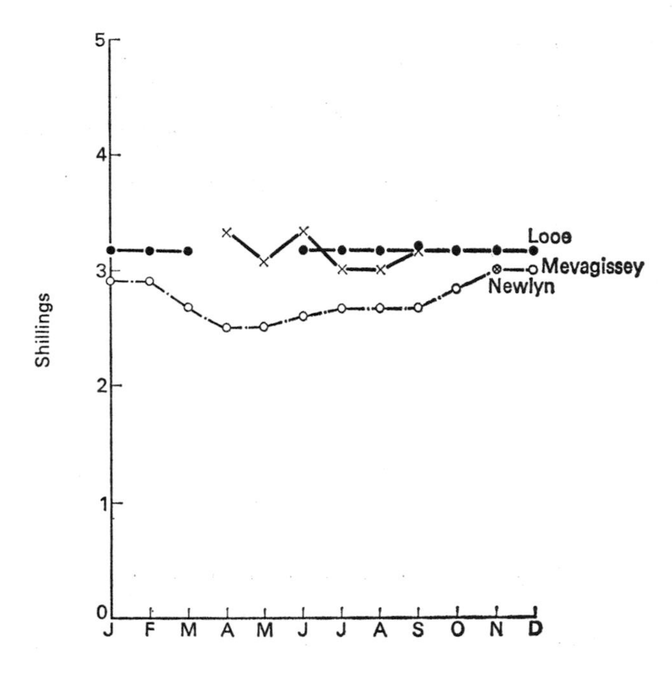

FIG. 13.8. Monthly prices of pilchards per stone at Newlyn, Looe and Mevagissey, 1961.

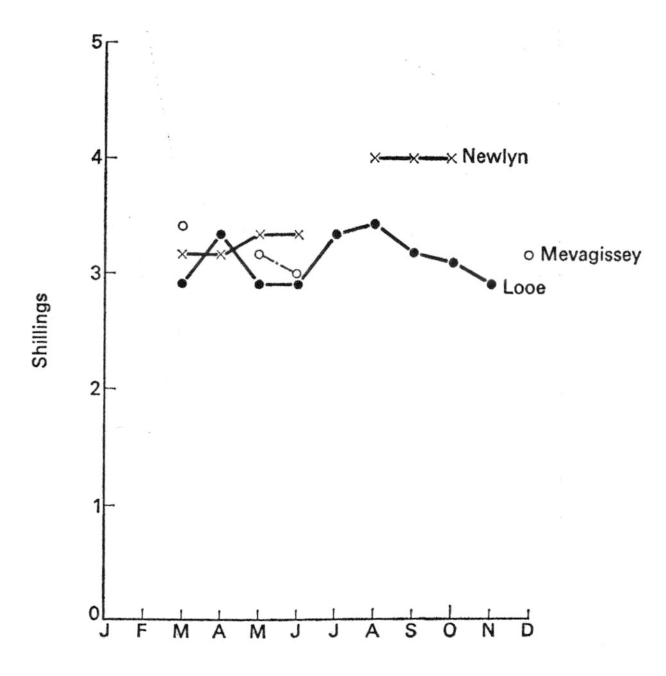

FIG. 13.9. Monthly prices of pilchards per stone at Newlyn, Looe and Mevagissey, 1966.

In 1951 (Fig. 13.6) the Newlyn prices did not seem to bear very much relation to those paid in Looe and Mevagissey as they were higher than in these two ports from April to November, but they fell below both in December. There is no apparent explanation for this sudden December fall as many more fish were caught at the other two ports, only 3000 stones being landed at Newlyn. A possible explanation is that the Newlyn canners did not require many more fish and had therefore lowered their prices. In fact the price per stone of pilchards on first landing at Newlyn in December 1951 had dropped to 3*s.* 8*d.* per stone, which was lower than any price since June of that year. In Looe and Mevagissey the prices were very close together throughout 1951 and, on the whole, rose during the year.

In 1956 (Fig. 13.7) the three ports were all paying higher prices than in 1951. The average price of pilchards on first landing for the whole of that year reached the highest figure for

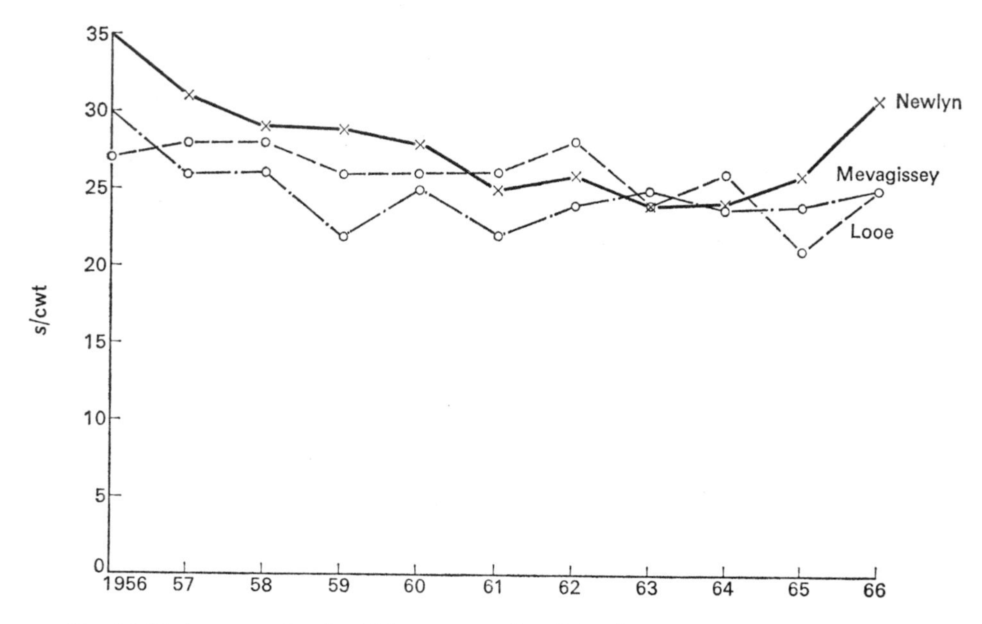

FIG. 13.10. Average prices in shillings per cwt for pilchards at Newlyn, Looe and Mevagissey, 1956 through 1966.

the 16-year period from 1951 through 1966 (see Fig. 13.5). As mentioned earlier, however, there was more agreement on the actual trend in prices, as prices at Newlyn and Looe tended to fall from May to October and approach the fixed price of Mevagissey.

In 1961 there was very little difference between the prices obtained in the three ports. However, that year was significant in that the average price paid to the fishermen at Newlyn for pilchards, was below that paid at Looe (Fig. 13.8). Figure 13.9 shows the situation concerning prices which prevailed in 1966. Although prices in all three ports fluctuated, those at Newlyn were steadier and, except in March and April, higher than elsewhere. The prices at Looe fluctuated somewhat but fell steadily from August through November. The spasmodic appearance of figures for Mevagissey suggests that the closing of the curing plant there had severely affected the fishing from the port.

From 1957 through 1962 the price paid at Mevagissey was the lowest of the three ports (Fig. 13.10). This is possibly because at Mevagissey the price paid by the salting firm would be below that paid by the canners, and because the proportion of salted to canned fish was probably greater there than at Newlyn; this would have the effect of lowering the overall price of the fish at the port. The prices at Newlyn have tended to be higher than at Looe or Mevagissey, but between 1961 and 1964 they were below those at one or other of the other two ports. This was the result of a falling trend since 1956; for example, in 1956 the average price obtained per stone of pilchards was 4*s*. 4*d*. at Newlyn but by 1961 this had fallen to 3*s*. 1*d*. (Figs. 13.7, 13.8). There was also a similar downward trend at the other two ports although prices fluctuated at Mevagissey and were more stable at Looe. After the closing of the Looe canning factory to pilchards in 1962 Mevagissey was placed in a stronger position than Looe and prices obtained there came more into line with those obtained at the other two ports. The variations in prices obtained at Newlyn and Looe from 1963 through 1966 have been considerable, so that no clear trend can be established; Newlyn prices, however, have risen since 1964.

To summarize, Newlyn remains the major port of the pilchard industry, but after the war until the mid-1960s became relatively less important because of the processing facilities available at Looe and Mevagissey. If these continue to decrease then Newlyn will become the only place where pilchards will find an assured market and it will then be the centre of the industry.

CHAPTER 14

PROFITABILITY

PROFIT is not an easy subject to define, and before doing so it is useful to introduce the idea of the entrepreneur. An entrepreneur can be thought of as a person who combines factors of production to the greatest advantage in the light of changing economic conditions. Originally this was applied to individual businessmen who were risking their own capital and were responsible for taking decisions as a result of which the capital would be utilized in the most advantageous way. The application of the term in this original sense to modern enterprises is rather difficult as the people who make the decisions nowadays are not usually risking their own capital. Nevertheless, as Seldon and Pennance (1965) point out, the concept of the entrepreneur is still very important in economic thought. In the context of this chapter its relevance is enhanced because it introduces the notion of uncertainty in industry which is particularly pertinent to the fishing industry. In fact, the very existence of profit is dependent on risk-taking and can be considered as a premium gained, or lost, as a result of risk-taking. Profit can now be defined as the cost of entrepreneurship. Looking at it in another way it can be said that capital invested in a concern must be paid for, and this is done using the residual money obtained from sales after all other costs have been deducted. Alternatively, the rate of profit per annum on capital used in industry can theoretically be defined as, "the net yield (after deducting depreciation charges) expressed as a percentage of the current cost of the assets" (Benham, 1955).

The return on capital invested in industry should be above the rate of interest found, for example, in Government Stock. This is considered necessary in order that capital can be attracted to a concern where the element of risk involved is greater than elsewhere. If no risks were involved in capital investment in industry then rate of profit and rate of interest would be equal. There are certain industries, however, and certain branches of the inshore fishing industry fall into this category, in which lower returns on capital are accepted for compensations of a non-monetary nature. In the West Country of England at least, fishing is not only a business it is a way of life. That is, the men engaged in fishing for pilchards in Cornwall are prepared to receive less actual profit from their capital invested in the industry because it enables them to live the kind of life which they enjoy. Thus, capital invested in the pilchard industry is providing them with a benefit, known as a social benefit, which is non-calculable in balance-sheet terms. It could be argued that these men are also of political value as they could provide part of the Navy of the country in the event of war. Again, it is not possible to calculate the value of this aspect of fisheries, so that the net profitability of the fishing industry of Cornwall will always remain rather conjectural. Similar arguments can also be applied to some other branches of the fishing industry; but in the larger commercial fishing enterprises where many investors are involved, then the company concerned must provide realistic returns on the investors' capital in order to attract investment. It then becomes impossible to look upon considerations of the subjective nature discussed above as adequate substitutes for interest.

In the next few pages the profitability of the industry is considered. The subject is dealt with under three headings, namely primary, secondary and tertiary profitability, which are defined *in situ*. The treatment has been deliberately kept general, with some points of particular interest being added where possible, so that the approach to the subject can be more clearly seen.

Primary Profitability

First, the costs and earnings of the individual units must be assessed. In this case, the unit is considered to consist of the vessel, engine, crew, nets and other necessary items of equipment. The following elements are regarded as parts of the total costs of primary fishing enterprises; that is the processes involved in bringing the fish from the sea to the place of first sale, but not including any sales commissions on the fish. This is included by some authorities, but if primary concerns end with the fish being landed at the dockside, then no sales commission can be included.

1. Depreciation.
2. Labour (i.e. crew's wages).
3. Maintenance.
4. Insurance.
5. Fuel and lubricants.
6. Landing costs.
7. Overheads of shore establishments.
8. Managerial remuneration.
9. Entrepreneurial remuneration.
10. Miscellaneous.

These will now be dealt with in turn in a little more detail.

1. *Depreciation*

In considering depreciation it is found that there are three main components of this factor. These are the depreciation of

(i) the hull,
(ii) the engine, and
(iii) other equipment such as nets.

Depreciation is difficult to assess as no uniform method of calculating it has been adopted by the fishing industry. According to Holliman and Ovenden (1959) the following methods of recording depreciation have been used.

(a) The straight line method. In this the whole cost of the vessel is written off in ten, fifteen or twenty equal instalments.
(b) The diminishing balance method in which 10 or 15 per cent is written off each year. This is based on the current theoretical value of the vessel at the end of each year.
(c) A notional method in which the value is written off possibly at the same rates as those allowed for taxation.

(d) When dealing with several vessels a certain amount is taken from profit to allow for the depreciation of all the vessels.
(e) Provision for depreciation may not be made, and this is particularly prevalent among inshore fishermen.

The annual amount of depreciation of the value of the three main items mentioned above is not necessarily the figure included in the concern's commercial accounts or even that allowed for taxation purposes. The White Fish Authority in estimating depreciation use two methods. Thus, for steel motor vessels or oil-fired steam vessels it is taken as the lesser of either 10 per cent of the insured value of the hull and machinery on 31 October for the year being considered, or $6\frac{2}{3}$ per cent of the purchase price. For other vessels it is taken as 10 per cent of the insured value of the hull and machinery.

In the pilchard industry the boats are very old and so in fact will, in most cases, have reached the final point of their depreciation. Further, as this is a particularly conservative branch of the inshore fishery, it is likely that no provision for any depreciation is made in their costings, although those that do depreciate their vessels do so on a basis of 5 or 10 per cent each year (method (b)).

2. *Crew's Wages*

This can be considered to consist of both fixed and variable cost factors in many cases. That is there may be a fixed basic wage paid whether trips are made or not, or whether they are profitable or not. This also includes national insurance payments. Then over and above this is the variable element, the share money, which will depend on the man's position and the success of the trip.

In the pilchard industry, however, the situation is again different as there is no statutory wage, remuneration being entirely dependent on the share system. Thus, this cost is completely variable. An example of how the share system might operate in Cornwall is given below but the system is very flexible. The profits of a vessel might be divided into eight shares. The gross takings first have the costs of the particular trip deducted, and this covers such items as fuel, bait if used, landing dues, possibly food and any other item of a similar nature. If each member of a crew of four receives one-eighth of the remaining takings (known as one body share), this leaves four-eighths which are allocated to cover costs of repairs and maintenance to nets, engine and other gear and also to the purchasing of any item of equipment that may be necessary. This latter item may mean only a temporary allocation of a share to this purpose. For example, if an echo-sounder is being purchased a share may be given over to this for the duration of the purchasing proceedings.

In some ports, instead of half the shares being allocated non-specifically to engine and gear, either or both of these items may have a share allocated solely to them. In some cases investment by persons other than the skipper is made in a vessel or its gear and the investor then receives his share from the catch. This practice is not common.

There are many variations of the system and these are found, not only from port to port, but also from vessel to vessel within a port. According to one authority (March, 1955), during the nineteenth century the share system was more standardized in that the catch was divided into eight shares which were allocated as follows. One went to the boat, three to the nets, and four to the men. The skipper received the same share as the other members of the crew, and the boy was allowed the value of the fish which fell into the sea during hauling of the nets. Incidentally, these fish were retrieved using a net on the end of a long pole and known as the kieve net.

3. *Maintenance*

There are certain items of maintenance which always have to be done however much the vessel is used. These items of fixed cost include keeping the hull painted, an operation usually done twice a year on pilchard drifters, and keeping the engine in good running order. However, the number of trips in any one season will determine to a large extent how much has to be spent on repairs to both the hull and the engine and other equipment. This part of the cost is, therefore, variable. The cost of paint and repairs to the hull comes out of the shares allocated to the boat, but the skipper pays for the engine repairs and maintenance.

4. *Insurance*

When considered over a short time period, insurance costs are fixed. Nevertheless, over a longer period, due to depreciation, no-claims bonuses or alteration of premiums in some other way, insurance can also be considered as a variable cost.

All owners of pilchard drifters in Cornwall have their vessels insured by a local Fishermen's Co-operative System or Mutual Society, which enables the premium to be low. It is 2 per cent of the value of the vessel and engine per annum, and each year there is a rebate, the size of which depends on the number of claims the society has met during the year. However, it is usually in the region of 10*s*. per 100*s*. of the premium. Personal accident insurance schemes are carried by some vessels and these guarantee payments for specified accidents, and a sickness benefit for about £7 per week for 52 weeks (Great Britain, 1963a).

5. *Fuel and Lubricants*

Depending on how many trips are made and of what duration, this is, of course, a strictly variable cost.

All the Cornish pilchard drifters now have diesel engines which use about three gallons of fuel per hour. Diesel fuel is relatively cheaper for use in agriculture and fisheries than for ordinary domestic consumption.

The cost of lubricants is also very variable but appears to be a very small item for the pilchard industry (Bridger, personal communication, 1963).

6. *Landing Costs*

Several elements of costing are included under this item which includes harbour and dock dues, the cost of labour to unload the catch, and certain miscellaneous items such as boxes and ice. Conventionally rental of radio and echo-sounders, etc., is also included in this item (Holliman and Ovenden, 1959). Obviously, it contains elements of both fixed and variable costs, and some of these items will not be directly relevant to the pilchard industry. Also where this fishery differs from larger fisheries, is that here the skippers are usually the owners of their vessel and own only one vessel. Manufacturers of equipment such as echo-sounders will not hire sets to individual owners, but only to companies.

Actual landing dues in Cornwall vary from port to port, but are based on the value of fish landed, probably averaging out at about threepence in the pound. Similarly dock dues are variable but are rated at so much per foot, based on the registered length of the vessel. However, neither of these charges is very high.

7. *Overheads of Shore Establishments*

This item includes those costs connected with shore establishments concerned with the management of vessels only. In large concerns such as are found along the east coast these costs are considerable. However, in the pilchard industry shore management is done in the home, and costs connected with shore establishments as such would not be a factor that can be considered.

8. *Managerial remuneration*

The manager of a large concern naturally receives payment for his services. In a consideration of the costs of the pilchard industry, where the manager is usually the skipper of the vessel, it should be remembered that the skipper's wages must be included in the costs.

9. *Entrepreneurial Remuneration*

The entrepreneur of the pilchard fishing unit is, again, usually the skipper. His capacity here as an investor, however, is entirely separate from his managerial capacity. He should, in theory, receive some payment for the use of capital risked in the enterprise and this should be a percentage of the profit. If he is the sole investor he should receive the whole of the profit and this necessitates considering profit as a cost. This cost is bound to be very variable, however, and in practice may not always be positive.

To present some indication of the relative importance of the various items of cost the following balance sheet is quoted from official sources (Great Britain, 1963b). (The costs are estimated as percentages of the whole balance and Table 22 is intended to give a general, quantitative picture of the factors involved.) It gives typical costs for a pilchard drifter

TABLE 22. BALANCE SHEET FOR PILCHARD DRIFTER IN CORNWALL

INCOME	%
1. Total value of fish sales	88·2
2. Subsidy	9·4
3. Other income from boat	2·4
	100·0

Expenses as pilchard drifter		Expenses as long-liner	
1. Labour (includes National Insurance and working owner's share)	47·2	1. Labour	47·2
2. Maintenance		2. Maintenance: (a)	2·6
(a) Vessel and engine repairs	2·6	(b)	4·5
(b) Fishing gear	4·5	3. Insurance	1·7
3. Insurance	1·7	4. Fuel and lubricants	4·9
4. Fuel and lubricants	4·9	5. Landing charges	1·6
5. Landing charges	1·6	6. Hire charges	0·5
6. Hire charges	0·5	7. Bait and baiting	15·3
7. Miscellaneous	6·1	8. Ice	1·0
		9. Food	1·2
		10. Miscellaneous	6·1
Total expenses	69·1	Total expenses	86·6
Operating surplus (entrepreneurial remuneration)	30·9	Operating surplus (entrepreneurial remuneration)	13·4

which may be used during some of the year for long-lining. It is for this latter method of fishing that the costs of ice and bait and baiting are included.

Having given the relative importance of the various expenses it is now proposed to give some idea of the actual size of the various figures. To this end the following estimate of the income of a pilchard drifter was made.

One W.F.A. officer gave the average catch of the pilchard drifter to be 7000–14,000 stone p.a. The price obtained varies from about 2*s*. 9*d*. to 3*s*. 3*d*. per stone. If the total catch per boat is taken as 10,500 stone p.a., and the price per stone as 4*s*. (including the subsidy until August 1966 when subsidy payments changed), then the average income from pilchard drifting is about £2100.

The official statistics, as quoted in Table 22, then suggest that about another £50 p.a. (2·4 per cent) is earned by the boat from sources other than pilchard drifting. However, it is well known that for some months of the year pilchard boats go out long-lining, and a rough estimate of the earnings of pilchard boats long-lining from Looe was made. It is possible that another £500 can be added to the income figure bringing the total now to about £2600.

The figure can be left here, but other sources of income, the size of which is difficult to estimate, can be mentioned. Shark fishing, in which about six holiday-makers are taken out for the day at a cost to them of 30*s*. to £2 per head, brings in about £7 10*s*. 0*d*. per day for several weeks; a figure arrived at as shown in Table 23.

TABLE 23. INCOME FROM SHARK FISHING PER DAY

Item	£	*s*.	*d*.
6 passengers at 30*s*. per head	9	0	0
Fuel—about 3 gal/hr at 1*s*. 3*d*. per gal for 8 hr	1	8	0
Total income from this source per day	7	12	0

It can be seen that during the holiday season of 8–10 weeks, three or four trips a week would add another £200 to the income from the boat.

Also, over and above the normal fishing operations, feathering for mackerel in season generally takes place. Income from this source would depend on the intensity of the fish shoals and upon the market price, but a crew member can add £20 a week to his earnings in a good season for a short period at the beginning of the mackerel fishery without too much difficulty.

It can be seen, therefore, that the official estimate of the income of a pilchard drifter could easily be an underestimate of £1000 p.a.

With reference to the incomes of the individual crew members, the official W.F.A. statistics state that for a man working on a 40-ft vessel fishing with drift nets and lines, the average earnings are £422 p.a. including National Insurance contributions. The earnings of the skipper are dependent on the arrangement of the share system for individual vessels. It is almost certain that any estimate of the income of pilchard fishermen, based on official sources, will be considerably lower than the actual income. This is due to their supplementary sources of income mentioned above, which cannot have an official check kept on them. These are of especial "value" also as they are not likely to be fully taxed.

Returning now to the profitability of the industry as a whole, elements of costing beyond those concerned with bringing the fish to the dockside must be considered.

Secondary Profitability

The movements of the fish from its handling at the ports to its subsequent first marketing will now be dealt with. These movements are said to constitute secondary profitability. The market at which the fish arrives at the end of this stage could be the final market, or it might be that the product is then made available to wholesalers or export merchants.

The costs at the secondary stage of production depend on the nature of the final product. That is, in the case of the pilchard, they depend on whether the fish is to be sold fresh, salted or canned. The factors of cost involved in the different processes will now be considered.

1. *Fresh Fish*

The pilchard is rarely sold fresh, but when it is the secondary costs involved are: (i) salesman's commission, (ii) transport from dock to retailer.

2. *Salted Product*

The following items have to be considered when the pilchard is produced in the cured form:

(i) Salesman's commission.
(ii) Transport from dock to curer.
(iii) Cost of raw materials—fish and salt.
(iv) Labour.
(v) Overheads concerned with depreciation of machinery and cost of buildings, etc.
(vi) Barrels.
(vii) Others.

Balanced against these costs are the proceeds from any by-product which may be saleable as a result of the manufacture of the main product. In this case there might be a small profit from the sale of the oil which is pressed from the fish.

3. *Canned Product*

When the pilchard is canned, the following costs (similar to those stated above) are involved:

(i) Cost of raw material.
(ii) Transport to canner.
(iii) Labour.
(iv) Overheads.
(v) Fuel.
(vi) Cans.
(vii) Others.

Tertiary Profitability

This refers to those elements of the industry concerned with handling of the product between the factory and the consumer. The costs include: (i) Transport from producer to (a) wholesaler, or (b) retailer.

If an intermediate wholesaler is involved then (ii) profit of wholesaler, (iii) transport from wholesaler to retailer, will also be included in this category.

It has not been possible to obtain any official figures† but an essay at costing for the cured product was made and this is given below:

(This estimate includes secondary items of cost as well as tertiary.)

Basic price pilchards per stone	3*s.* 0*d.*
Cost of cask per stone	1*s.* 3*d.*
Price of storage in vat and processing per stone	1*s.* 0*d.*
Transport to harbour per stone	1*s.* 3*d.*
Labour for loading on to ship per stone	6*d.*
Packaging per stone	1*s.* 0*d.*
Unloading at Ostend per stone	6*d.*
Transport Ostend to Italy per stone	2*s.* 0*d.*
On arrival in Italy	10*s.* 6*d.* per stone

† I am, however, indebted to J. P. Bridger for discussion of this subject, upon which these figures are based.

Thus, on arrival in Italy the product costs 10*s.* 6*d.* per stone. The price may be slightly lower than this if any oil obtained from the process is sold, and the proceeds from this balanced against these costs. Although at first sight the increase in price of the product has been very large, even allowing for retailer profit in Italy, the product should not finally cost more than 10*d.* to 1*s.* per pound weight. For a food of the nutritional value of salted pilchard (see Chapter 12) this is still very reasonable. Although it has been suggested that the fall in Italian sales has been due to large increases in price of the product since the war, due to rising transport costs, it is much more likely, as mentioned before, that the fall in sales is due to the increased standard of living in Italy during the past two decades.

When canning firms were approached they would not furnish any details which would be of use in costing the product. However, the *Report of the Committee of Inquiry into the Fishing Industry* states: "The actual cost of the fish forms a relatively small part of the selling price of a tin of pilchards—the first-hand cost of the fish in a 14-oz tin retailing at 1·9 to 1·11 pence at the date of our enquiry (1960–61) is between 2*d* and 3*d* and even if heavier supplies led to a large fall in the price of fish, it would make comparatively little difference to the price of the final product."

"We are doubtful", the report continues, "whether a much expanded industry would be able to compete with imported pilchards backed, as they are, by plentiful supplies of both fish and cheap labour" (Great Britain, 1961). In the 7 years since the date of that report the situation has not altered significantly and the majority of canned pilchards sold in Great Britain are still African in origin.

The Subsidy

It has been stated that examination of costs and earnings of a fishery are of "prime importance in fisheries run-down economically in which government loans or assistance in some form is required" (Stolting and Murray, 1958). The studies of costs and earnings help to determine where and how subsidies should be given. The government inquiry into the fishing industry (Great Britain, 1961) suggested that the subsidy on all vessels should be changed from the system whereby it was calculated on quantity landed to a method of payment per day at sea. This payment would be calculated from the gross earnings of various classes of vessel over a given period. The report, however, admits that "a daily rate of subsidy may be impracticable for certain inshore vessels" and for these the Committee suggested that "the subsidy should continue to be paid on the quantity of fish landed, and the rate being adjusted to correspond as closely as possible to the percentage of average gross annual earnings fixed for larger vessels".

This was borne out by another authority (F.A.O., 1958) which stated that information collected on costs and earnings serves two main purposes in relation to the subsidy:

1. It assists in assessing the amount of subsidy to be spent on each section of the fleet.
2. Subsidy rates, being related to the size of the vessel, can be apportioned correctly to the different length groups, after the collected data have been studied.

The subsidy for the pilchard has been paid to the fishermen at the rates of about 1*s*. per stone for several years. However, the price paid for ungutted and gutted fish has always been different. For example, for the year between 1 August 1965 and 31 July 1966 the subsidy on ungutted fish was 10½*d*. per stone and on gutted fish 1*s*. 1½*d*. per stone. After August 1966, however, the organization of the subsidy for the inshore fleet was altered so that vessels between 35 and 60 ft in length were paid a flat rate of so much per day at sea provided that they had been paid at least £500 in stonage in the previous year. The rates for the year 1 August 1966 to 31 July 1967 and 1 August 1967 to 31 July 1968 were as set out in Table 24.

TABLE 24. RATES OF SUBSIDY FOR INSHORE VESSELS PER DAY AT SEA

(Figures from Great Britain, 1967, and Personal Communication, 1968)

Length of vessel (feet)	Subsidy 1966/67	Subsidy 1967/68
	£ *s*. *d*.	£ *s*. *d*.
35 or over but under 40	3 15 0	3 10 0
40 or over but under 45	4 0 0	3 15 0
45 or over but under 55	4 10 0	4 4 0
55 or over but under 60	4 15 0	4 9 0

Those vessels whose subsidy during the year 1 August 1965 to 31 July 1966 did not amount to £500 were paid at the rate of 1*s*. per stone for gutted fish and 9½*d*. per stone for ungutted fish during the year 1966–7. For the year 1967–8 the figures were 11*d*. and 8½*d*. for gutted and ungutted fish respectively.

The subsidy has been one of the major factors contributing to the continuance of the industry, but it is doubtful whether the information collected on costs and earnings in the pilchard has had the desired effect in relation to the distribution of the subsidy. That is, it is not immediately obvious that the subsidy is benefiting only the fishermen. In order to be able to compete with African imports the price of pilchards per stone paid by the canners to the fishermen could not rise above a certain upper limit. With the subsidy at the present rate, it is now bringing the pilchard fisherman a quarter of his income from pilchard drifting which places the subsidy in an unusual position in relation to the industry as a whole.

Thus, although the subsidy as paid to the fisherman is initially designed to enable only his side of the industry to operate at a profit, in effect it is possibly acting as a subsidy also to the processing side of the industry, especially the canners. The present position from the canners' point of view is that if the price per stone were to rise too much their product would be priced out of the market and the foreign imports would take over the market completely. Also from the canners' point of view the subsidy is useful in that the fishermen are content with a lower price per stone from them when the price is made up by the government. With the situation as it stands at present, it is certain that the fishermen could not manage if they were paid a lesser price per stone than they are at present, also it is probable that the canners could not compete on the market if they had to pay more per stone for the fish. It seems, therefore, that the Fleck Committee was correct when it stated, as quoted earlier, that it seemed "doubtful whether a much expanded industry would be able to compete with imported pilchards, backed, as they are, by plentiful supplies of both fish and cheap labour" (Great Britain, 1961).

It should be remembered, however, that the largest single exporter of canned pilchards into this country is South West Africa which with South Africa in 1963 and 1964 exported 11,735 and 12,843 tons of canned pilchards to the United Kingdom, which was nearly 10 per cent of the total imports of the product for those 2 years. At present (1968), although no longer in the Commonwealth, South Africa still enjoys Imperial preference, being exempt from import restrictions and this also includes goods imported from South West Africa.

For non-Commonwealth countries there is a 10 per cent *ad valorem* tariff on canned pilchards being brought into the United Kingdom. That is, the country's importers of these goods have to pay what amounts to a tax of 10 per cent of the value of the import price of the goods. In April 1963 a government paper was published announcing the terms of the Anglo-Japanese Commercial Treaty which was to come into operation on 4 May 1963 (Great Britain, 1963). This freed any Japanese canned pilchards coming into this country from import licensing restrictions, but now practically no canned pilchards come to the United Kingdom from this source.

If now some restriction, either tariff or quota, were placed on the South African imports, then depending upon which restriction was imposed, there would be one of two resulting changes both of which would be advantageous to the Cornish pilchard industry. If a tariff were imposed on them then the market price would rise up to a point at which consumers would change to home products. If there were a quota restriction the imports would be limited with, however, a similar ultimate effect.

As long as the Cornish pilchard has the cheaper South African product as a direct competitor it will only be able to take over the whole market by using one of the two methods mentioned above. If, however, it could branch out into other lines, then the foreign goods would no longer be direct competitors, and new markets for the Cornish products could be secured without the necessity of political measures being taken. Some suggestions for new products have already been made and they will be further discussed in the next chapter. Unfortunately, it is at present impracticable to talk in terms of the Cornish industry taking over the marketing niche occupied by African products because the catching side of the industry is now so run down that it would need a major capital outlay for the industry even to begin to move towards the quantity of landings required. In view of the huge demand for canned pilchards in Great Britain (14,482 metric tons in 1966—Great Britain, 1967) it would be unwise to erect any form of trading barrier against these African products until the English industry is more able to fulfil the country's demand.

CHAPTER 15

GENERAL CONCLUSIONS CONCERNING THE ENGLISH PILCHARD INDUSTRY

AFTER the war economic and biological conditions were ideal for general expansion in fisheries. Economically, because money was being invested in a newly expanding industry; biologically, because the stocks had been able to recover from the effects of fishing between the wars. This was particularly noticeable in the North Sea stocks, individuals of given age groups of which were shown to have increased in size when compared with pre-war sizes (Fig. 15.1). In fact expansion also occurred in the pilchard industry, but this did not include structural integration such as took place in the east coast industry.

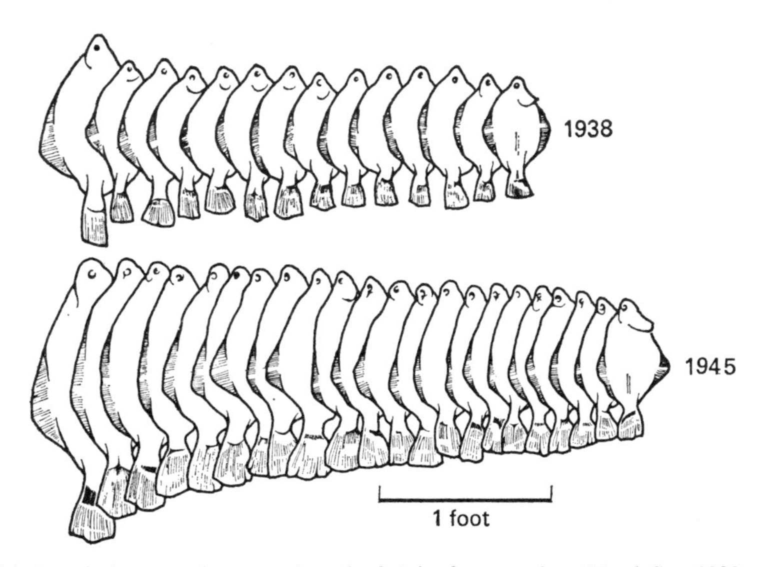

FIG. 15.1. A typical quarter-hour trawl catch of plaice from southern North Sea, 1938 and 1945.

As mentioned earlier (Chapter 7) there was a general upward trend in the size of the annual pilchard catch from 1946 to 1951, and it was during this time that there was an expansion on the processing side of the industry. Up to 1955, there were nine canneries regularly taking supplies of fish from the area (Kennea, 1957), six of these were in Cornwall or Devon and the other three were as widely scattered as Chichester, Leytonstone and Great Yarmouth. By 1960 only three remained in the West Country (Tysser, 1960), and the Leytonstone factory no longer operated. In 1962 another of the canners closed down, and this one had been very important to the pilchard industry, so reducing the total of canners concerned with the pilchard industry to four. At the present time (1968) no canning firms have any connection with the industry except one at Newlyn and one at Chichester. In the latter case the main utilization is in processing other than canning. The large number of canners in the early 1950s was the result of an attempt by certain firms to take advantage of a somewhat expanding industry and yet, with the catching side of the industry as it was, it must have been possible to foresee that such an expansion could not last.

The catching side of the industry has already been described in detail and it was mentioned that the drift fishery has been in existence since the early years of the seventeenth century. In fact, since that time there have been no fundamental changes in the drift fishery The size of the vessels used and the length and mesh of the nets have not basically altered. The industry was thus not capable of coping with a consistent demand such as arose when the nine canning factories were operating. An industry, the catching side of which was geared to domestic, or, at the most, manually operated commercial preserving methods, could not be expected adequately to supply the demands of mechanized processing techniques. The gradual closing down of the factories was thus inevitable. Unfortunately, at the same time, the export trade for cured pilchard began to decline and by 1956 was no more than a few hundred tons per annum. This led to the situation that still prevails and which is described below.

The normal path of distribution of pilchards is to the canners by prearranged contract. The two or three firms interested in the catch for processing are sufficient to be able to dispose of the normal landings, and they are in a very powerful position. In fact, they control the size of the catch of the drifters. This is because when they have received a series of good catches from the local drifters they soon reach the limit of their capacity. They then inform the fishermen that no more fish are required at that time. The amount of fishing is thus governed largely by the size of the throughput of the processing firms. Two quotations from very different sources will illustrate this point. The first is from *Fishing News* (1962a) and states: "because the firm could not take any more fish for the time being, pilchard boats at Mevagissey have been stopped from fishing by the Duchy Canneries Ltd." (This firm is no longer concerned with pilchards.) The second quote, from a White Fish Authority Report (Great Britain, 1958), states: "the year was a difficult one. . . . Owing to the accumulation of stocks at the beginning of the year, canners considerably reduced their purchases from the fishermen during the spring and summer season."

A few more fish can usually be sold after the canners have reached their limit, and these are sent to one of the two remaining curing plants in the area. However, this is only a very small outlet, although curing was formerly the main standby during the time of gluts, and when the curers can take no more fish any excess has to be dumped back in the sea. The paradoxical situation thus prevails of there being too few canneries taking pilchards at certain times, and yet too many for them all to be constantly supplied with pilchards.

This might lead to the conclusion that it is the processing side of the industry which is too poorly developed and which has prevented a normal development of the catching side. This, of course, is the fisherman's point of view. On the other hand, it is equally legitimate to argue that, with the catching side at its present primitive stage of development it would be unwise to expand the processing side, especially when the history of the pilchard canneries in the area is considered. Furthermore, it is argued that any development of processing techniques is absolutely dependent on a regular supply of pilchards and this should be assured before any other development of the industry as a whole could even be considered.

It has been stated, most emphatically, by the representatives of the fishermen, that it was the marketing side of the industry which was at fault. Whether they mean by this that the markets for the existing pilchard products are not fully exploited and need developing, or that research should be directed towards developing new markets and new products, is not made really clear. Certainly, by 1956–7 circumstances had conspired to produce a very unhealthy state in the Cornish pilchard industry which can only be described as a vicious circle.

The situation was then, and unfortunately remains today, that the canners will not expand their possible canning capacity for pilchards because they say the supply is too irregular and in any case too small; the fishermen say they will not increase their catching capacity because the canners do not always take all that they catch at present, and there are no satisfactory alternative markets.

All these complaints concerning the state of the pilchard industry indicate that if it is to revive and develop, research has to be done to determine where the "bottlenecks" in the industry may be removed; where the vicious circle could be broken. It is also essential to know whether the basic biological resource could survive any larger permanent exploitation.

This is the fundamental question, because even though the Cornish fishermen always claim that if the fish are required they can catch them, it seems certain, for two reasons, that this is an optimistic viewpoint. First, it is unlikely that seven canning factories would have had to close down or stop taking pilchards if the fishermen were supplying them constantly with the raw materials they required. Secondly, if the industry were to expand noticeably the annual catch would have to rise very considerably above the figure of 2000 tons which has been the maximum catch recently.

In 1957 the report by Cushing (1957), already referred to, made an estimate that the stock of pilchards in the Channel was probably in the region of 800,000 tons. This report was based on work done in 1949 and 1950, and it seems probable that a considerable increase in the stock of pilchard had taken place compared with the 1920s and earlier. Thus, according to Southward (1963), "It would appear that clupeoid fish off Plymouth are now at least twice as abundant as they were before 1930 to '35, and possibly very much more than twice." As was mentioned previously in Chapter 3, lack of data makes quantitative estimates difficult (and Southward mentioned this obstacle when he stated: "we cannot properly measure the various biological changes, partly because of lack of earlier quantitative evidence, and partly because of present difficulties in obtaining quantitive evidence." Nevertheless, even if Cushing's estimate were as much as 30 per cent optimistic, the pilchard catch could still be increased considerably with the standing stock of about 500,000 tons to which such an error would reduce it.

Evidence accumulated by the Marine Laboratory at Plymouth (Southward, personal communication) suggests that spring-spawning pilchard (May to July) have been much less common in the Eddystone area since 1963. There seems to have been a corresponding increase in the Autumn spawning fish, as shown both by egg counts and by occasional trawl catches of adult fish. The herring population has increased during the same period and there have been changes in the plankton, but it is too soon yet to be certain if there has been a permanent alteration in the stocks of pelagic fish.

However, with Cushing's estimate as a guide, the White Fish Authority suggested (Great Britain, 1959) that stable conditions in the industry could not be obtained until the main pilchard resources further out in the Channel were exploited: "Only thus could regular and sufficient supplies be assured to the canneries at prices which would enable them to operate on an economic basis." The report then went on to recommend that a pilot scheme, using a vessel equipped for different methods of fishing, should be put into operation. Besides determining the extent of the resource, this would also be able "to provide data on potential costs and earnings and on the type and requirements of future craft to be employed in the industry".

By the time that a later W.F.A. report had been published (Great Britain, 1960), the recommendations of the early reports were becoming more definite and the objectives of a research project (which was eventually to become called the Pilchard Development Unit) were briefly defined as "to ascertain the availability of pilchards in fishing areas within reach of the Cornish ports and the most economical means of catching them". At the same time, technical and economic studies of processing and marketing were to be made. During the early part of 1960 a vessel was chosen, which would be converted for the research, and a committee was called to supervise the work. The W.F.A., the Ministry of Agriculture, Fisheries and Food, the Cornwall Sea Fisheries Committee, the Food Manufacturers' Federation and the fishermen themselves were all represented on this committee.

In a special report (Great Britain, 1961a) it was stated that the committee had decided to charter the 65-ft diesel-powered trawler *Madeline* to study the fishery resource. In the report the authority admitted that some people thought that processing and marketing should be studied first, but justified their decision to start with research into the biological side because they deemed it "wise to confirm from the early results that there was a real prospect of heavier and more regular catches, and secondly, because it was felt that the availability of the additional supplies should be known within reasonable limits before the requirements of processors could be assessed".

After several setbacks the *Madeline* was eventually made ready for the research work and equipped to take various types of gear other than, but including, drift nets. (These included the Vinge trawl, midwater trawl and purse seine.) Also, as mentioned earlier, there was a wide range of electronic equipment on board. The electronic gear was used for locating and estimating the intensity of the pilchard shoals, and some of the results obtained from the *Madeline* have been shown in Figs. 4.2 and 4.4. The different catching gears were used experimentally to determine whether methods other than drifting might be used more economically and successfully to obtain the pilchard. In 1961 the report of the Fleck Committee said of the pilchard industry and the research programme associated with it:

> Pilchards are one of the few species of which large unexploited stocks are believed to exist within easy reach of our shores. The White Fish Authority, in consultation with local interests, has been investigating the possibility of building up a more extensive industry, and has made plans for the exploration of the offshore waters in search of more ample supplies, and for a technical and economic study of the shore side of the industry. [Great Britain, 1961.]

In early February 1963 the research was brought to a close and, after two additional research trips to the area in March and November and December 1964 a report was published which, to summarize, stated that plenty of fish were located by the echo-sounder, but that during the summer, due to the scattered nature of the shoals, drifting was the only method of catching them consistently. However, the mid-water trawl might possibly yield satisfactory results when the fish shoal more densely and closer inshore during the winter, although these would need to be operated from vessels between 100 and 200 ft in length. Even these vessels, however, would need drift nets in rougher weather as the shoals break up under such conditions (Bridger, 1965).

It can be seen that a number of factors such as the declining catches since 1958, the cheaper price and wider availability of imported pilchards, the increasing irregularity with which the fish have been caught in Cornish offshore waters which probably gave rise to the reluctance of canners to take pilchards as a main line, the decline of the salting industry and its associated export trade, and the unwillingness to adopt any new method of dealing with gluts have led to the pilchard industry being referred to occasionally in the press as a dying industry (*Fishing News*, 1962). Yet, due to recent research, it is possible to say how some of these factors causing the decline of the industry could be ameliorated, given the necessary capital and receptiveness to innovations in the West Country. Some of these possible future developments are summarized in the following pages.

Before the pilchard industry can expand it is necessary to show that the fish not only are available, but that they are being caught. Pilchards have been notorious throughout the history of the industry for their changeable habits. For example, they no longer come as close inshore in large shoals as they did in the past. They have in recent years been shoaling densely, however, during the winter months between Falmouth and Plymouth (Bridger, personal communication, 10 February 1963). Nevertheless, despite this knowledge, it has not been possible during the past 2 years for the drifters to make the comparatively larger catches which they should have been able to do, as the weather has been very bad. Yet one purse-seine shot of the *Madeline* of 409 stones, in November 1962, was made during this bad period, actually during a time when the remainder of the fleet was unable to operate due to adverse conditions (*Fishing News*, 1962b). If the drifters had been able to change to this alternative method of fishing as the *Madeline* could, then in all probability the catches for November 1962 would have been several times greater than they were. The *Madeline*, of course, is 20–30 ft longer than most of the drifters and this extra length is usually necessary for methods of fishing other than drifting. Auxillary gear such as a power winch is also essential, and trawling of any sort requires a higher power to weight ratio than is found on the conventional drifter.

It would seem, therefore, the pilchard drifter of 30–45 ft length is no longer going to be a profitable proposition. Such a vessel can only drift safely within 20 miles of the coast and the majority of its time it is well within these limits about 5–10 miles from the shore. A vessel of greater length would be capable of carrying a winch and have the power required for trawling and would also be able to use other methods of fishing. It would also be able to continue drift-net fishing, but this would be possible further out at sea. Then the pilchard resources which, the recent research confirms, are there during the summer months, though scattered, would be made more available to it.

The method of suspension of the net in the water would probably need to be altered if summer drifting in deeper waters further out to sea were to become common practice. This is because the pilchard net is a sunk net (see Chapter 8) and has no heavy warp at its base. This enables the net to hang vertically only in calm weather. Heavy swell and strong currents cause it to be suspended at an angle away from the vertical and reduce its catching area. The herring net, on the other hand, is a swum net and is stabilized in the water by the heavy tarred warp or messenger rope attached to the bottom of the nets. If the pilchard drifters were to adopt this method of suspending their nets (and the change would not entail a very great cost or alteration of technique), then the pilchard drift net would become a more efficient catching method, less dependent on the weather.

Drift nets have traditionally been made of cotton, but recently various synthetic materials have been tried in their manufacture. Having overcome the difficulty of slipping knots the synthetic nets are now claimed to be as efficient as the cotton nets, with the advantage that they do not need treating with cutch as do the cotton nets, but with the disadvantage that they are said to have a tendency to cut the fish. The *First Progress Bulletin of the Pilchard Industry Development Project*, already referred to, says of the nylon nets tried by the *Madeline* that they "caught about the same quantity of pilchards per unit length of net and the fish were in good condition" (Great Britain, 1961a). The report goes on to say that eventually it will be possible "to determine whether in the long run nylon nets are more economical than cotton". Viewed in the light of the fact that the Cornish pilchard fishermen mend and remend their nets until it is impossible to tell which is the original net and which has been added, a practice which makes it very difficult to assess the annual outlay on nets, this statement by the W.F.A. is perhaps rather optimistic. The following state of affairs is much more likely to exist in effect. Although about one-fifth of a fleet of twenty-five nets would need replacing each year which, if the nets were bought new would cost about £230, in fact they tend to be obtained second-hand. This, and the extensive mending of the nets, probably means that expenditure per annum on nets is no more than about £50. It would thus be difficult to estimate which of the two materials would be the more economic for the purpose. Experiments have been carried out with different coloured nylon nets as it was thought that certain colours might be less easily avoided by the fish than others. However, there appeared to be no significant difference between the catches of the different nets.

It was mentioned earlier (in Chapter 8) that a reduction in the mesh size from 36 to 42 rows per yard might give a better yield of smaller fish. These smaller fish, apart from commercial considerations of the greater popularity of sardines over pilchards, are of greater nutritional value than the larger ones, and this change might benefit the industry in several ways. If it were decided that an alteration from the sunk to the swum pattern drift nets would be advantageous to the industry, it would be worth considering that a change to the smaller mesh size might also give further advantages to the drift-net fishery.

Whilst a reduction in mesh size and change in the method of suspension of the drift net would increase the efficiency of drifting, this could possibly also be further increased by mechanizing the method of hauling. A method of using the Puretic power block system in the hauling of drift nets has been devised by Puretic and Schmidt and was described as follows:

> Basically, the net would be hauled in the same manner as at present, using the cable around the winch with a modified power block supported by the boom across the vessel from the hauling side. The power block would take the strain on the net, which would be controlled by a foot pedal or knee-actuated lever, operated by a man located in the normal position for shaking out the fish. The net would come across the gunwale, where two men would pull it apart, shaking out a small amount of the catch at a time. [Schmidt, 1959.]

Before any process is changed from a manual method of operation to a mechanized one, there are certain considerations of an economic nature to which attention must be given. Installation of a machine involves several factors, some of which are disadvantageous. Disadvantages, which can be translated into monetary terms may be:

1. The initial cost of the machine may be too high. It must be decided whether or not it would be possible to repay this cost (with interest) in the estimated life of the machine and whether there would also be enough saving compared with the methods to cover especially;
2. the cost of extra fuel necessary and
3. the cost of maintenance.

The advantages can be represented mainly by the following points:

1. *Labour*. (a) Direct saving of costs in crew number and, therefore, wages. It is always a difficult problem to decide on automation, especially in an area where unemployment can

easily occur; and Cornwall is one of these areas. (b) Indirect saving of costs of administration, etc.

2. *Greater flexibility*. This point need not necessarily apply to all mechanization processes but, for example, concerning the further mechanization of the canning process discussed in Chapter 9, it was mentioned that a machine which fed the fish into the cans would enable the factory to work throughout the night without the "disproportionate increase in expenditure which is inherent when human labour is necessary, due to overtime rates".

It is also possible that certain human labour which is released by mechanization of one process can be more fruitfully employed in another position, and this could be particularly marked on board a fishing vessel.

3. *Greater dependability*. When a machine is designed to carry out a job which has previously been done by human labour, at the start of the job at a given time the efficiency of the two methods might be equal. However, the speed of completion of the job is usually several times greater by the machine, and added to this is the advantage that a machine is not subject to the normal human tendencies of reducing speed and efficiency with time. All these factors have to be weighed up before a final decision to introduce mechanization is made.

A further mechanization is possible in the drifter. This is the mechanization of the shaking of the fish from the nets which is possible and a machine for this purpose has been developed by the Russians, and this reduces the time required for shaking herring from a drift net to 1–3½ minutes (Fishing News (Books) Ltd., 1964).

As for the other methods of fishing that have been tried in the West Country, most of them have had only moderate success and are useful only at the times of the year when the fish are shoaling densely.

For example, mid-water trawling met with success provided that there was not much *Noctiluca*, which causes phosphorescence, in the water. Experiments aboard the *Chichester Lass* and the *Madeline* have yielded results which compared favourably with those of drifters fishing at the same time and in the same area, especially when it is recalled that the nets and boards were being modified during the experiments. An echo-sounder is essential for mid-water trawling, as warps have to be paid out to bring the trawl to the same depth as the shoal, and a headline oscillator is another piece of electronic equipment which is now considered to be almost essential for success to be achieved in this method. This is attached to the headline of the mouth of the trawl, and indicates the depth of the trawl in the water. The actual ratio of "length of warp paid out: depth of trawl in the water" is approximately 5:1.

In 1947, experiments with the ring-net method of fishing aboard the motor vessels *Onaway* and *Hope* proved to be successful when the fish were shoaling, and this method of fishing has many advantages over the drift-net method (Hodgson and Richardson, 1949). Hodgson and Richardson considered that certain Cornish pilchard boats were suitable for conversion to ring netters, but the method has the disadvantage that two boats are required to work together, which is probably why the method has never been tried since. (The subject has been dealt with in more detail in Chapter 8.) Vinge trawling and fly seining both yielded completely negative results and purse-seining, as mentioned above, was successful but only during the winter.

In order to make full use of the resource it appears that a pilchard vessel needs to be at least between 60 and 70 ft in length and to be equipped with power winches, a fleet of swum drift nets, and possibly a set of gear for an alternative method of fishing. This latter should be capable of catching fish when they are shoaling densely and, from the evidence that has arisen from research in the area, a mid-water trawl or purse-seine would be best suited to the purpose.

Most skippers of pilchard vessels at present also own gear for long-lining. This method of fishing is used when pilchards are in bad condition or short supply, and when it is thought that the species caught by the lines will fetch a good price on the market. This alternative should be retained even if larger pilchard vessels are built. With the larger vessels, greater lengths of line could be set further out to sea (as is done by the east-coast long-liners) and the pilchard fishermen would then still be able to carry on a lucrative fishing operation at a time when pilchards were not required by the canning factories because of their lean condition.

It has been shown that pilchards are present in large quantities in the Western Channel. These fish can be caught but not in consistently large quantities by the pilchard fishermen at present using a conventional size drifter and drift net. One of the hazards of all fishing enterprises is the fluctuations which occur in the catches, but in the pilchard industry these have been exceedingly marked. The change from salting to canning, as the chief method of preserving the fish, has meant that these fluctuations have suddenly become serious in that they affect the overheads and labour recruitment problems in a cannery much more greatly than in a curing works.

Consistent catches could be supplied if larger pilchard drifters equipped as mentioned above were built for use in the West Country. The grant and loan system of the W.F.A. would, of course, have to be used by fishermen wishing to buy the larger boats and equipment necessary. Possibly a smaller, technical subsidy should be made available for fishermen wishing to make their vessels more efficient by mechanization of some part of the fishing procedure. It is essential that grants and loans and subsidies, when made, should be channelled in the correct amounts to various branches of an industry. The W.F.A. have, for some time, been conducting inquiries into the costs and earnings of inshore fishermen and it is possible that from these they would be able to determine where and how the grants and loans and/or subsidies should be given. Current costs have to be balanced against prospective costs and earnings and in this way, as stated by Stolting and Murray (1958), new methods should not be adopted "without adequate consideration being given to the cost of discarding present equipment, to the impact of these innovations on other channels of the production processes, and to capital investment".

It was stated earlier that a vicious circle exists today in the pilchard industry. If this were to be broken by an increase in the number of fish caught due to new vessels and equipment being used in the area, then it is likely that the processing side of the industry would not be able to deal with the full quantity of fish caught. However, once canning firms had been shown that the fish were being consistently caught, then more investment in this direction would be made. In the interim period gluts would undoubtedly occur and it should be possible for arrangements to be made for those fish not cured, to be treated at a fish meal factory. Although there is no factory of this kind in the West Country and transport of fish to the east coast is not an economic proposition for this purpose, it would be only a short-term policy until the processing methods had become geared to the new rate of supply. Recently an application has been made to the White Fish Authority for financial aid to establish a fish processing factory in central Cornwall (*Fishing News*, 1965), and such a development would be a great asset to the British pilchard industry.

It is possible that gluts could be dealt with in other ways, and some research has been done on the problem. A summary of the work (which was complementary to that done in South Africa mentioned in Chapter 9) states that pilchards can be kept in good condition for canning for from 3 to 4 days if they are well iced when caught and then kept in ice. Although longer periods of preservation of pilchards, of up to 6 months, can be obtained by quick-freezing the fresh fish and then storing them at −20°F, this hardly seems either practicable, because of the expense, or necessary because of the fact that 3–4 days is usually long enough for canning production to even out. It seems, therefore, that the problem of gluts has been exaggerated. The main problem is to persuade the fishermen to take ice to sea with them.

The argument against this has been that drifters are too small for it to be practicable to take up the available space with either ice or boxes. Some have taken boxes with them when the incentive of a larger price per stone was offered for boxed fish, and it is possbile that room could be found for ice in a similar way. However, in the future, if the drifters were built 70 ft or more in length, trips of longer duration would be made and then both ice and boxes would have to be carried to enable the fish to retain their freshness.

Pilchards can be preserved in brine after landing and held in large containers and then steamed under pressure. Unfortunately this process tends to break up the fish, because of the length of time which the steaming takes, and they are then only useful for processing to fish pastes. However, fish pastes are a useful outlet for this product and possibly give a greater margin of profit to the manufacturer than other methods of processing pilchards used at the present time. It is, perhaps, a product which should be given more attention in both the variety of flavours combined with the basic fish, and in the way in which it is presented to the public. The traditional paste jar now has a dull and conservative image. Paste sold in a tube or attractive plastic container could well have more consumer appeal and at the same time be a cheaper method of packaging the product.

Another possible use for pilchards is in the manufacture of fish fingers or fish sticks. These fall into the category known as breaded fishery products. The manufacture of such goods has been steadily increasing since the mid-1950s because of their convenience in domestic preparation. In addition, to commercial users they provide the following advantages:

(i) uniform servings which allow
(ii) easy cost control;
(iii) small storage space and ease of handling;
(iv) rapid preparation;
(v) lower labour costs may result from (iii) and (iv).

The manufacture of breaded products consists of five stages which are briefly outlined below.

1. Preparation of the raw material. The most important aspect here is the cutting of the portions to a given, uniform weight and size from the frozen raw material.
2. Coating the portions with the batter which is now usually done mechanically.
3. The breading of the portions with a mixture designed to give an attractive golden-brown colour when cooked.
4. Frying. This is a very rapid process, the coated and breaded portions being dipped in oil at about 200°C for from 20–45 sec.
5. Packaging and freezing which is usually also followed by an inspection for defects in quality, appearance and weight (Learson, 1969).

The use of pilchards in fish fingers would help to make pilchards much more frequently bought for human consumption as many purchasers would not realise that the fish fingers contained pilchard. Similarly, due to the rather anonymous flavour of the fish frequently incorporated in this type of product, due perhaps to the spicing which is also used in their manufacture, people with a vague and ill-founded objection to pilchards as human food would be persuaded (perhaps unknowingly!) to enjoy one of the richest protein foods available.

One final process which might be able to make use of pilchards when gluts occur is the manufacture of pilchard oil. This is possibly only a small outlet and details of quantities used in this process are not available, but there is at least one firm in the north of England which manufactures this product mainly for use as bait. The fish are crushed in water and the oil is then skimmed off. Heating may also be necessary and the oil is then refined over a range of temperatures depending on the quality of oil required (Thomas Aspinall Ltd., personal communication, 31 May 1963).

When more fish began to be caught regularly for at least 9 months of the year, probably the most useful investment on the processing side would take the form of a factory able to produce pilchards in a variety of guises. Such a factory would be able to take a large quantity of fish and regulate its production of different lines as the condition of the fish or the market demanded. Although managers of processing factories disagree, it would seem essential that pilchard products should be more varied than the normal canned fish in tomato sauce. Their argument against any innovation is that pilchards in tomato sauce is what the public demand. This is, of course, all that the public is offered, and suitable advertising of a new variety would soon persuade people to try pilchards in a different form. The more likely reason why new products have not been tried is that pilchards have never been a particularly profitable product from the processors' point of view and any extra expense involved in the production process would initially lower their profit margin even further. This seems to be rather a short-sighted view.

At present the tomato sauce product is generally recognized as a cheap product and therefore, if the classification of fish from Chapter 12 is recalled, the product is in the ordinary class of fish. Special sauces, such as those which have been used by German processors in some of their canned herring products, and a newly designed pack and label, could raise the "status" of pilchards. This could also be further enhanced by a simultaneous production of the marinated product. These products would be more expensive than the traditional pilchard pack or the imported product, but with the rise in consumption of convenience foods, mentioned earlier, provided that the release of these products onto the market was correlated with an advertising campaign, the chances of their selling well are very high.

Obviously such a plan for the revival of the pilchard industry involves a great deal of capital expenditure, and half measures with an industry which is at such a backward stage of development would not achieve the desired result. Once the new products were selling well and there was a regular throughput in the factories, which require as regular a supply as is possible with a basic resource such as fish, then a natural expansion and development of the industry can be envisaged. This would also entail a movement of younger men into the industry, and a certain amount of structural integration, both vertical and horizontal. Such development would alter the basic character of the industry as it is today. It would become less of a "vocational" industry, and the skipper–owner would tend to be replaced by larger operational units due to the integration. Fishermen would then not consider entering the pilchard industry solely because they like the life, as they do today, and they would no longer be prepared to accept less monetary returns on their capital because of the subjective returns of leading the sort of life they like best. However, if the Government were to lend substantial money to the industry, then any plans for its development should be in the national interest. That is, any development plan should aim at decreasing the parochial nature of the industry and increasing its contribution to the national economy and world food production.

PART III

THE SARDINE INDUSTRY OF CALIFORNIA

CHAPTER 16

INTRODUCTION, HISTORY AND THE RESOURCE

W. F. THOMPSON, writing in 1926, stated:

> There has been a truly marvellous development of the sardine fishery in California. Although it originated as a great fishery during the stress of war, the industry has shown a vitality which augurs well for its permanence so long as the raw material is obtainable. The amount caught exceeds by far that taken of any other species in California, and there appears at present no other which is capable of the tremendous yield, unless it be the unused anchovy.
>
> Experience with older fisheries has shown that rational use demands a knowledge of at least two things. There must, above all else, be information from time to time regarding the manner in which the species is withstanding the strain of the fishery. There must also be an understanding of the natural changes in abundance which inevitably occur, so that these may be distinguished from the effects of overfishing and also may be foretold and understood. Based on such knowledge, regulation and exploitation may be rational and restrained.

This quotation indicates the manner in which the sardine industry of California should have developed and at the same time it underlines some of the ways in which the management of the fishery has not been successful. Throughout its history the industry has been the subject of concentrated and consistent research which initially makes the downfall of the fishery very difficult to understand. The quotation above was written at a time when investigations which were to cover all fisheries of the State of California were being initiated, and shows that those scientists working in the State had an early insight into the difficulties of rational exploitation of a resource, especially a resource like the sardine, which by its behaviour of massive shoaling bares itself to mishandling through overfishing. In the account of the California sardine industry which follows, it will be shown how a very rapid expansion occurred which was followed by an equally rapid decline, and that in spite of forecasts of the stock being in danger, accompanied by recommendations to reduce the amount of fishing effort being expended, short-sighted commercial policies with considerable power over-rode these other considerations. The result was that severe overfishing occurred so that from a peak year for landings in 1936–7 there was a slump in the mid-1940s, and in spite of slight occasional improvements the industry has not since been able to recover its former buoyancy.

The development of the sardine industry in California is comparatively recent, the first references to fishing for sardines on the Pacific coast of America being for 1889. From then until 1892 practically the whole of the catch was landed in San Francisco County, the only place where a pilchard cannery was operating at the time. This cannery had started at San Francisco in 1889. According to Dewberry (1954) canning had been first introduced to the Pacific coast of the United States in 1864. This was for salmon, the menhaden being the first "sardine"-type fish to be canned in the United States around 1870. The catch in the vicinity of San Francisco, however, was uncertain and so the cannery closed and reopened

TABLE 25. LANDINGS OF PACIFIC SARDINE ALONG THE COASTS OF CALIFORNIA, OREGON, WASHINGTON AND BRITISH COLUMBIA

(Metric tons; figures for 1916–17 to 1951–2 converted from Clark, 1952; figures for 1952–3 to 1964–5 from F.A.O. Statistics; figures for 1965–6 and 1966–7 converted from California, 1967; 1968.)

Season	California	Oregon	Washington	British Columbia	Grand total	California per cent of total
1916–17	24,970				24,970	100
17–18	65,850			70	65,920	99
18–19	68,520			3,300	71,820	95
19–20	60,810			2,980	63,790	95
1920–21	34,880			3,540	38,420	90
21–22	33,110			900	34,010	97
22–23	59,070			930	60,000	98
23–24	76,150			880	77,030	99
24–25	157,000			1,240	158,240	99
1925–26	124,510			14,470	138,980	90
26–27	138,060			43,540	181,600	76
27–28	169,850			62,070	231,920	73
28–29	230,900			73,020	303,920	76
29–30	294,940			78,310	373,250	79
1930–31	[a]167,910			68,090	236,000	71
31–32	[a]149,340			66,760	216,100	69
32–33	[a]227,370			40,230	267,600	85
33–34	[a]351,780			3,670	355,450	99
34–35	[a]540,210			39,000	579,210	93
1935–36	[a]508,390	23,790	9	41,100	573,289	89
36–37	[a]640,760	12,880	5,950	40,320	699,910	92
37–38	[a]377,830	15,110	15,510	43,610	452,060	84
38–39	[a]521,780	15,440	24,020	46,960	561,240	86
39–40	491,840	20,250	16,110	5,010	533,210	92
1940–41	417,760	2,870	730	26,100	447,460	93
41–42	532,760	14,380	15,510	54,470	617,120	86
42–43	457,786	1,770	530	59,760	519,846	88
43–44	433,680	1,650	9,470	80,490	525,290	83
44–45	502,410		18	53,620	556,048	90
1945–46	366,100	80	2,100	31,110	399,390	92
46–47	211,880	3,590	5,570	3,620	224,660	94
47–48	110,050	6,290	1,230	440	118,010	93
48–49	169,300	4,830	50		174,180	97
49–50	306,190				306,190	100
1950–51	320,260				320,260	100
51–52	114,980				114,980	100
52–53	6,500				6,500	100
53–54	4,300				4,300	100
54–55	61,900				61,900	100
1955–56	66,000				66,000	100
56–57	31,500				31,500	100
57–58	20,800				20,800	100
58–59	94,100				94,100	100
59–60	33,700				33,700	100
1960–61	26,100				26,100	100
61–62	19,600				19,600	100
62–63	7,000				7,000	100
63–64	3,200				3,200	100
64–65	6,000				6,000	100
1965–66	890				890	100
66–67	<700				<700	100

[a] Includes quantity landed at floating plants.

at San Pedro in December 1893 (W. L. Scofield, mimeographed notes). Another cannery was also opened at San Pedro, so that by 1895 the industry had gained some momentum there. Fishermen in the San Francisco area, deprived of their cannery, and those in other places too far from San Pedro usually managed to dispose of their sardines as bait for rock fish.

The packing of sardines in cans about 1890 made the United States the largest producer of canned fish at that time (van den Broek, C.J.H., Fish Canning, in Borgstrom, 1965).

In 1909 a cannery started operating 150 miles further south at San Diego and took small quantities of sardines. However, a regular supply of fish was eventually maintained by the introduction of the lampara net, about 1907, and this method was soon adopted along the length of the California coast. This enabled canneries to be reopened in the more northern district of Monterey. Prior to this, catches had been made with gill-nets or beach seines. About 1929 the purse-seine was introduced, an advance in the evolution of catching techniques for shoaling fish that had to be preceded by the development of the power winch. This combination of gear enabled the boats to make larger catches than had hitherto been possible.

It was in 1916, as a result of various stimuli provided by the First World War (dealt with in more detail later), that the industry really began to reach sizeable proportions in California. Thus the catch in the 1916–17 season had been 24,970 tons. The total catch from California fluctuated, but from 1916 until the 1936–7 season there was an overall tendency to increase. The peak was reached in 1936–7 with a catch of 640,762 tons from California alone and another 59,165 tons from areas bordering the Pacific, north of California. Since that time there has been an overall decline until the present. The landings in 1966 amounted to less than 1000 tons (California, 1968).

Catches of the sardine elsewhere along the Pacific coast of North America and Canada have been made but are very much smaller than the California catch (as shown in Table 25 from Clark, 1952) and do not greatly affect considerations that follow, nor are they likely to alter any conclusions which might be made. Thus, for the 36 years between 1916–17 and 1951–2 California caught 90 per cent or more of the total landings of the Pacific sardine for 21 of these years, that is 58 per cent of the time. This is not to underestimate the important part the landings from these northern areas have played in the research programme designed to investigate the biology of the sardine. Indeed they have been important in many ways as, for example, in unravelling some of the patterns of migration. Since 1951–2 considerable quantities of sardines have been caught by fishermen from Baja California. Figures for these landings from 1951–2 through 1964–5 are given in Table 26. (I am indebted to Dr. F. N. Clark for sending me these.)

The Biological Resource

The sardine industry of California is based on a single species of fish known biologically as *Sardinops caerulea* at present, although it seems likely, as discussed in Chapter 2, that within the not too distant future it will become known as a sub-species of *Sardinops sagax*.

Its distribution is from the tip of Lower or Baja California, and into the Gulf of California, in the south, to south-eastern Alaska in the north. These are the limits of its range and its distribution is open to variation within these limits owing to environmental variations. Its northern appearances have always been seasonal and since the decline of the fishery its appearances off the Canadian coast have been much reduced, in both the frequency and size of shoals.

TABLE 26. LANDINGS OF PACIFIC SARDINE IN BAJA CALIFORNIA, 1951–2 THROUGH 1964–5

Season	Quantity
1951–52	14,450
52–53	8,180
53–54	12,770
54–55	11,170
1955–56	3,760
56–57	12,050
57–58	8,860
58–59	19,940
59–60	11,610
1960–61	17,770
61–62	18,990
62–63	13,900
63–64	16,590
64–65	24,110

Feeding

The feeding habits of the Pacific sardine are similar to those of the Cornish pilchard. That is to say, it is a plankton filter-feeder assimilating both phyto- and zooplankton with preferences for each at different times of the year and of the life history.

As reported by Ahlstrom (1960), Arthur (1956) in his doctoral thesis found that the food of the larvae consists of copepod material as eggs, or at the nauplii or copepodid stages. Feeding at this stage is selective and particulate as the gill rakers do not develop until metamorphosis.

TABLE 27. FOOD OF LARVAE OF *Sardinops caerulea* UP TO 25·00 mm LENGTH (After Ahlstrom, 1960)

Food taken	% taken by larvae up to 5·5 mm length	% taken by larvae from 6·00 to 9·5 mm length	% taken by larvae 10·0 to 25·00 mm length
Copepod eggs	22·0	10·8	28·6
Copepod nauplii	54·5	76·3	37·1
Copepodid stages	0·3	3·4	31·4
Miscellaneous	4·4	—	—
Unidentified material	18·8	9·5	2·9
	100·0	100·0	100·0

Table 27 from Ahlstrom (1960) summarizes the food of the larvae. Hart (1937) stated that copepods remain the chief item of food until they are about 7 or 8 cm in length and this has since been confirmed by Arthur (1956), who said that the food of the larvae of the Pacific sardine is less than 1 per cent plant material. In 1949 a survey of the food of the sardine was undertaken by Hand and Berner (1959). Of 571 fish examined, crustaceans contributed 89 per cent of the organic matter in the stomachs; 74 per cent of the total was actually small copepods. During the summer the larger fish of about 10·0 cm length seemed to have diatoms as their most important food with crustaceans now second in importance (Hart, 1937). This is a state of affairs very similar to the Cornish pilchard, and, as will be shown later, to the South African pilchard.

It is not yet finally determined whether the Pacific sardine is selective in its feeding. Work on the Cornish pilchard suggested that, as individual items of the stomach contents were not present in the same relative proportions as in the sea, it was somewhat selective in its feeding.

Breeding

Spawning takes place during spring and early summer, with a peak in April or May. Reports of the depth at which spawning occurs vary, but it was observed (Clark and Marr, 1955) at depths down to about 20·8 fathoms (125 ft), and Ahlstrom (1960) gives the lower limit as 54·7 fathoms (328 ft). The temperature of the water is a critical factor, as it must be at least 12·5°C for spawning to occur, and not above 18°C. However, there have been some variations in the range quoted according to the results of individual workers. Hart (1937), for example, found no evidence of spawning where the water temperature was below 12·7°C, and judged the optimum range to be between 15° and 18·3°C. Clark (1940) suggested that the optimum range started at 14°C and extended to 18°C. Later it was thought (California, 1950) that the optimum range was from 12·5° to 16°C, as 98·4 per cent of all eggs which had been sampled had been taken within that 3½-degree range. Sardine shoals were found in waters having a wider temperature range from 11° to 20·4°C (California, 1950).

When in a spawning shoal their behaviour seems to be different from normal as they dart about excitedly, and frequently leap out of the water (Wolf, 1964). A normal sardine shoal appears crescent-shaped when seen from above, and this is the result of individual fish behaviour which MacGinitie and MacGinitie (1949) described as follows: "Sardines possess the interesting trait of orientating themselves so that they are a little behind and to one side of the fish that is ahead of them. This keeps them in a rather long and somewhat slender school." It might also be noted here that the tendency to form shoals has been observed to increase in water temperatures decreasing from 12° to 6°C and that the tendency is lessened by raising the temperature from 12° to 25°C (California, 1950).

The age at which spawning occurs has been difficult to determine, but Clark (1940) suggested that only a small percentage of the population spawns at the end of their second year, 50 per cent at the end of the third year, whilst at the end of the fourth year the majority spawns. It should be noted, however, that Clark's work was done when the fishery was in a much more flourishing state than it is at present, and still supporting catches of over 500,000 tons per season. Gates (1961), presumably referring to a more depleted stock, stated that the majority of the population spawns at the age of 2 and some spawn in their first year. This is also suggested by work of MacGregor's (1957) in which he found that all the female sardines he examined in 1946, from the San Pedro area, of 2 years of age and above had yolked ova, and 28 per cent of the 1-year-olds possessed yolked ova. This change in breeding age as illustrated by Clark's work and the two more recent examples has occurred simultaneously with the reduction in the life span of the sardine which has undergone curtailment since about 1928. Thus, as early as 1939 Clark showed that for the years 1918–28 the average life span was about 10 years; by 1933 this had been reduced to 6 years and by 1938 to 4 years. This has, of course, been reflected in the fishery as will be mentioned later, but it seems certain that this reduction is responsible for some far-reaching biological effects.

Apart from changing the relative proportions of the population breeding early it has

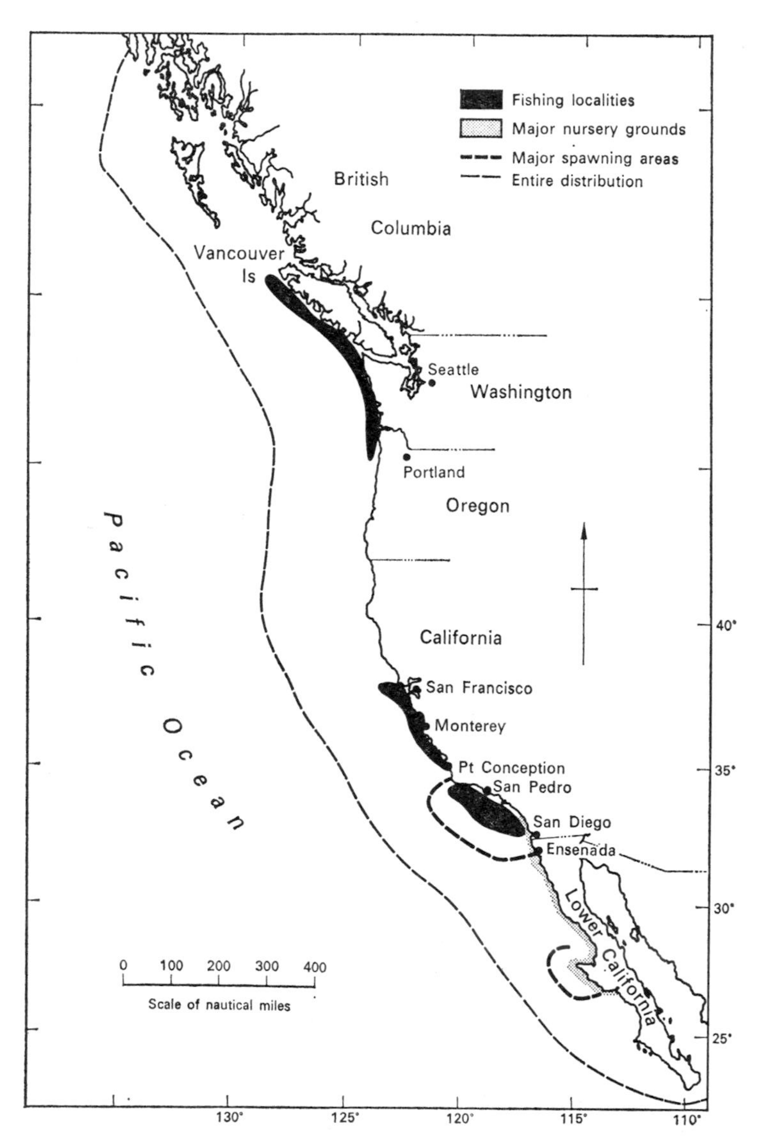

FIG. 16.1. Main spawning grounds and fishing localities off California and Lower California.

also affected the migration patterns and probably the balance of the fish populations in the waters of the Pacific off California. These points will be dealt with in more detail later.

From the extensive surveys which have been carried out by the California Co-operative Fisheries Investigations and earlier by the Fish and Wildlife Service of the United States it has been found that there are two main areas which are generally used as spawning grounds.

The major area is off southern California and northern Baja California between Point Conception and Ensenada. This is some 250 miles long and extends up to a distance of 200 miles from the shore. The other area, about half the size of the first, is off central Baja California. (Fig. 16.1.) There is also a third area, namely the Gulf of California, in which spawning occurs, but it is thought that these fish form a separate group and little is known about them at present (Ahlstrom, 1954; Gates, 1961). Ahlstrom (1954) and Marr (1957) thought that from these three areas it was possible to define four spawnings which are separated by either space or time. These are:

(i) the southern California offshore area. Here peak spawning takes place from April to May at temperatures between 13° and 16·5°C;
(ii) the lower California offshore area where peak spawning occurs from March to April between temperatures of 13° and 16·5°C;
(iii) the lower California offshore area with spawning later in the year from August to September at a higher temperature range of from 18° to 23°C;
(iv) the Gulf of California where there is probably a peak from February to March.

Apart from these main spawning areas thought to be favoured at present, before the collapse of the fishery sardine eggs had been taken off Oregon on occasions and there may also have been fairly regular spawning off central California (Ahlstrom, 1960).

The numbers of eggs spawned by female sardines is very variable. Small fish, of about 13–15 cm length may spawn about 30,000 eggs per season; larger specimens over 21 cm length may spawn up to 65,000 eggs per spawning. Estimation of the total number of eggs laid by any one female during a season is complicated, however, by the fact that the larger fish may spawn two or three times per season (Clark, 1934). On this reckoning a single female may spawn 200,000 eggs in one season, although some workers have recently questioned these figures (Gates, 1961). MacGregor (1957) was not certain from his experimental evidence whether there was more than one spawning, but it has not been possible to produce any more reliable estimates. This means that at present measurements of spawning stock size of the sardine based, as they must be, on an estimated number of spawnings, may be proved incorrect if MacGregor's contention, that there is possibly only one spawning per season, is correct.

Development

The egg, before and after fertilization, is very similar to that of the Cornish pilchard, measurements and external characteristics being quite similar for both. These characteristics may be summarized as follows:

(i) the total diameter is about 1·6 mm with a comparatively small yolk, although when just exposed to the sea water and before fertilization its diameter is slightly less, being about 1·1 mm;
(ii) there is a conspicuous oil globule;
(iii) there is a large perivitelline space.

These points are illustrated in Fig. 16.2. The eggs are deposited and fertilized in mid-water and, unlike those of the herring, remain pelagic. They are most frequently found at a depth of about 15 fathoms (Scofield, E.C., 1934; Hart, 1937).

The eggs take a few days to hatch, the exact time being dependent on the temperature of the sea. As mentioned previously, it has been shown (Ahlstrom, 1960) that at 12°, 14° and 16°C the eggs took 4, 3 and 2 days respectively to hatch. Ahlstrom (1943) has also shown that for each 0·5°C decrease in temperature the time until hatching occurs increases by about 7 per cent.

When hatching occurs a larva about 3.5 mm long is produced (Fig. 16.3) and it retains the yolk sac for 4–7 days before it is absorbed (Scofield, E.C., 1934). After 2 or 3 weeks it has grown to about 10 mm length. At about 34 mm (when it is about 2 or 3 months old) it becomes recognizable as a young sardine.

At the end of its first year a sardine is from about 14 to 19 cm long ($5\frac{1}{2}$–$7\frac{1}{2}$ in.), 2-year-old fish are from about 19–23 cm ($7\frac{1}{2}$–9 in.) length and 3-year-olds reach 25·5 cm, or 10 in.,

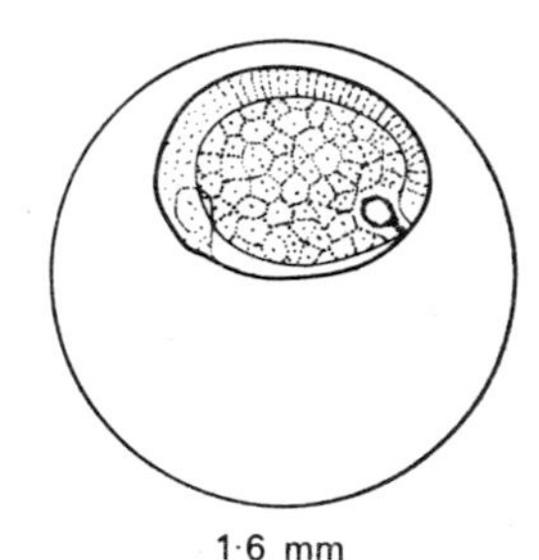

FIG. 16.2. Egg of Pacific sardine.

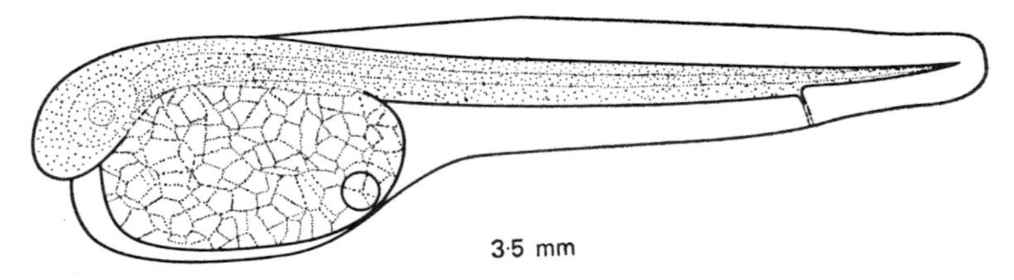

FIG. 16.3. 3·5-mm larva of Pacific sardine.

length. The maximum length achieved by fish as old as 10 or 12 years is not more than 30 or 31 cm, or about 12 in. This means that about half of the total growth takes place in the first year, the majority of the remainder occurring in the next 2 years with only minor increases in subsequent years.

Predators

The sardine has many predators, predators occurring at all stages of its life history. The eggs and larvae, both being planktonic are subject to the same predatory effects as other planktonic organisms of similar size. It is at those stages that the most serious effects of predation occur and Murphy (1961) suggested that it was the intense predation of plankton, including here the sardine eggs and larvae, that was likely to be "responsible for the rapid decline in numbers of the sardine larvae, and variations in this rate of predation are primarily responsible for variations in the rate of larval survival". From this statement it would appear that it is upon the extent of the depradations at these stages of the life history that the final size of the year-class (that is the number of fish of a given age group entering a fishery) ultimately depends.

Predators on the eggs and larvae include organisms from within the plankton such as copepods, chaetognaths and species of jelly fish and ctenophores. Non-planktonic organisms such as filter-feeding fish and including adult sardines also prey on these stages.

Juvenile and adult fish are preyed upon by a variety of organisms. Amongst the fish, tunas, sharks, yellowtail, barracuda, bonito, marlin, hake and mackerel are all known to be predators of the sardine; whilst mammals such as sea lions, seals, porpoises and whales, and birds such as gulls, pelicans and cormorants also prey on the juveniles and adults (Rosa and Laevastu, 1960; Gates, 1961).

Parasites

The full extent and effects of parasitism in fish generally are little understood. There are reports of parasites being found at most stages of development of the Pacific sardine. There are also reports of the occurrence of a fungus on the eggs (Gates, 1961) and bacteria have been reported at this stage (California, 1952).

Parasitic nematodes of the family Haemirudiae have been found to be present in the digestive tracts of most adult Pacific sardines (California, 1950) and parasitic trematodes have also been noted (Gates, 1961). Whether these have adverse or beneficial effects, however, is not yet known and much work would have to be done on this topic to determine this.

Shoaling

Clupeid fish, it is well known, have a marked tendency to form shoals or schools of fish. A shoal of fish has been defined (Blaxter and Holliday, 1963) as "a group of fish which are polarized and orientated in the same direction as a reaction to one another, rather than to a common reaction to an external stimulus. The fish tend to have a regular spacing and move at about the same speed."

Blaxter and Holliday (1963) have written a very interesting and useful review on shoaling. They pointed out that in some cases shoaling may have a survival value, as in *Sardinops caerulea* it had been observed that quite a small shoal being attacked by birds had formed itself into a sphere after which the birds had not been able to extract fish from it. Wolf (1964) confirmed that under attack sardine shoals tend to become spherical and that they bulge, indent and flow thus making the predators' sorties ineffectual. He reported that part of the sphere may be pushed right out of the water under heavy predation and that any fish which leaves the main sphere is generally captured. A recent article on the subject (Cushing and Harden Jones, 1968) also supports the theory that shoaling has some survival value under certain conditions, but when poor visibility prevails this advantage is lost.

To be weighed against this possible safety advantage is the disadvantage that the concentration of a species in the form of a shoal might attract predators. Furthermore, another possible disadvantage has been suggested by MacGregor (1959) who had proposed that shoaling might lead to poorer growth because of the density of individuals. On the other hand, Blaxter and Holliday argued that detection of food and hydrographical gradients by a shoal might be effective in bringing a greater proportion of the population into favourable conditions. This would operate on a differential effect whereby the part of the shoal in the most favourable environment would move slower than the remainder thus causing the other fish to swing round into the better conditions.

The general appearance of normal and spawning shoals of *Sardinops caerulea* has already been described. These shoals may contain very great numbers of fish, and Clark and Marr (1955) reported that there may be as many as 10 million fish in a shoal, "although one million or less is more common".

Migratory Movements

During their first year the young fish are moved more or less passively from the breeding grounds to a narrow strip of water along the shores of southern California and northern Baja California. There is a general inshore and southerly drift of currents in that area which effects this movement and results in the young fish being concentrated into the narrow inshore belt.

Movements of the adults seem to be very extensive within the geographical range defined above. These movements have been traced using a tagging programme which was carried out for a considerable number of years along the coasts of California, Oregon and British Columbia (Clark and Janssen, 1945a and 1945b; Janssen and Aplin, 1945). Evidence from these tagging experiments indicates that during their second summer, and in subsequent summers, fish move north from the spawning grounds off central Baja California and southern California. These northerly migrations are followed, during the autumn and winter, by a return to the south. The experiments have also shown that the older fish seem to travel further north, in their annual migrations, with increasing age, but that they always return to the south later in the year (Clark, 1940).

This pattern has become modified by the intensive exploitation of the stock so that the northward extent of the migrations is now curtailed. The older fish which, at one time, used to migrate as far north as Oregon and British Columbia from southern California, no longer exist in very great numbers. As a result of this the fishery off British Columbia which used to catch large fish with an average dominant age of just over 5 years (Marr, 1960) ceased after the 1947–8 season, as the fish were no longer present in the area in commercial quantities. The decline in the size of the landings during the last 5 years of the fishery at British Columbia is shown in Table 28.

At the same time that the British Columbia fishery was forced to close down, the most northerly sardine fishery in California, that in the San Francisco area, was almost brought to a standstill by an extremely low catch of about 2600 tons during the 1946–7 season. This

TABLE 28. DECLINE IN LANDINGS FROM BRITISH COLUMBIA FROM 1943–44 THROUGH 1947–48

(Figures from Marr, 1960)

Year	Catch (tons)
1943–4	80,490
1944–5	53,620
1945–6	31,100
1946–7	3,620
1947–8	440

bad season was followed by two much worse seasons when landings did not reach 100 tons. In spite of a slight revival during 1949–50 and 1950–1, the landings for 1951–2 were again below 100 tons and fishing for sardines in the area ceased.

Further down the coast at Monterey the landings followed a similar, but not quite so serious decline a year later. Thus, whereas in the 1946–7 season landings at San Francisco had been 2600 tons, at Monterey they totalled about 28,300 tons. The following season, however, with the San Francisco landings below 100 tons (actually only 81 tons) the Monterey figure had dropped to just under 16,000 tons. They continued at a low level until, during the 1951–2 season, under 14,500 tons were landed and the fishery ceased.

All this is evidence, admittedly circumstantial, that the migration patterns of the Pacific sardine have been altered by being severely curtailed due to the fishing effort. However, such theories have also to be considered in relation with other hypotheses such as that of Radovich (1962) who proposed a northern, genetically distinct, sub-population, and this is discussed in the next chapter.

CHAPTER 17

THE POPULATION IN RELATION TO THE FISHERY

For a resource to be properly managed it is necessary to know as much about it as possible. As Radovich (1962) pointed out, this includes finding answers to such difficult and much debated problems as that concerning the nature of the relationship between a year-class size and the size of the stock that spawned it. When a problem such as this is considered it is realized that much basic information is required before it can be solved. In this section a brief review of the information available on some of the more important topics will be given, following the plan laid down by Rosa (1965) for compiling information concerning the populations of aquatic species.

Structure of the Population

Sex Ratio

The sex ratio of the sardine population appears to be unity. That of the commercial catch has approached unity. Thus over a 15-year period (1941–2 through 1955–6) the percentage of females used for age determinations averaged 49·4 per cent (Ahlstrom, 1960). Certainly the seining method of catching used in California samples the population much more evenly than the drift net which the Cornish industry used and which, it was pointed out, gave a sex ratio which varies with the length of the fish caught.

Age

The age composition of the whole population has changed considerably over the last 50 years which is the period, of course, which covers the rise and decline of the industry. Thus, in the initial stages of the fishery the resource was under-exploited and the population could be expected to contain a certain percentage of individuals which would complete their natural life span. As the intensity of fishing increased, the number of individuals surviving for their full life span would be reduced; natural mortality would be reduced as mortality due to fishing increased. As this process continued the average life span has become reduced. Thus, as already mentioned, Clark (1939) reported that between 1918 and 1928 the average sardine life span was about 10 years, between 1928 and 1933 it fell to 6 years and between 1933 and 1938 it was reduced to 4 years. It can be seen, therefore, that an increase in fishing effort with a resultant increase in catch leads to increased fishing mortality and a reduction in the average age of the population, and the converse would also apply.

Age composition of the catch has been widely investigated, and reported on annually since 1949 (Felin *et al.*, 1950 and 1951; Felin *et al.*, 1952; Daugherty and Wolf, 1960 and 1964). A useful summary of the years 1932 through 1960 has been made by Wolf (1961a) who also presented the data in a graphical form (Wolf, 1961b). The average age of the catch is susceptible to modification by a single year-class. If it is relatively large when compared with those already in the fishery, it is generally referred to as a dominant year-class and will lower the average age of the catch as it enters the fishery. With successive seasons it will increase the average age until its size is reduced by natural mortality and by fishing so that its effect is no longer felt. Conversely, a relatively small year-class, generally referred to as a sub-dominant year-class, will increase average age. However, as it passes through the fishery its effect will become less and average age will decrease. Another general rule is that if for some reason there were to be no recruitment to the fishery then the average age of the population would, of course, increase.

A detailed survey of the age composition of the catch at San Francisco, Monterey, San Pedro and in British Columbia has been given by Marr (1960a). He demonstrated that in general, at all four regions, there had been a decline of the average age during the 1930's as would be expected by the increase in fishing effort which occurred during that period. This was the most important factor in determining the average age during that period. It appears that the fishing effort, by the end of the 1930's, had reduced the average age to such a level that fluctuations began to occur which were due to changes in year-class size as well as to changes in fishing effort. Marr pointed out, however, that at San Francisco changes in effort did not appear to be reflected in the age of the catch, but at Monterey and San Pedro it could account for 38 per cent and 24 per ccnt respectively of the changes in age composition. The mean average age for the four regions over a variable number of years for which records were available is shown in the Table 29 below:

TABLE 29. AVERAGE AGE OF SARDINES IN COMMERCIAL CATCHES OF FOUR AREAS

(Data over a number of years)

Area	Mean average age (years)
San Pedro	Slightly > 3
Monterey	Slightly > 3
San Francisco	Slightly < 3
British Columbia	5·5

The age at which the sardine becomes fully vulnerable to the fishery (known as the age of full recruitment) has been found to differ at various areas along the coast. Generally speaking the ports located in more southern areas report lower ages of full recruitment than those more northerly areas. Thus, the dominant age of full recruitment at San Pedro was 2 years, at Monterey and San Francisco 3 years and in British Columbia 4 years. It is likely that this can be interpreted in the light of the known migratory patterns of the Pacific sardine already described.

Length

As with age, so with considerations of the length of the fish in the population, it is very difficult to determine what average lengths appertain at various ages. This is because, with the population being considered, whose geographical range may stretch along about 2500 miles of coast, samples vary from region to region. However, taking the same regions that were considered for the age-samples it can be seen how the mean average length at these regions increases as their location becomes more northward. These are shown in Table 30.

TABLE 30. AVERAGE LENGTHS OF SARDINES IN COMMERCIAL CATCHES OF FOUR AREAS

Area	Mean standard length (mm)
San Pedro	211·9
Monterey	212·8
San Francisco	217·4
British Columbia	243·0

Two hypotheses have been proposed to account for this. The first suggests that the older the fish become the further north they reach in their summer migrations. As mentioned above, this thesis gains support from the evidence of age data, which clearly showed that fish from the British Columbia fishery were always of greater age than those from the California ports. However, Marr (1960a) whilst allowing that the sort of differential migration referred to does occur, suggests that it cannot account for all the differences observed, if average length for a given age is considered. The hypothesis put forward by Marr is that there are at least two groups of fish with different growth curves represented in the population. One group is represented by the fish from the British Columbia fishery and the other by those from the San Pedro fishery, whilst the similarity of those from Monterey and San Francisco suggests that these might be homogeneous mixtures of the two groups. The length-on-age curves given by Marr (1960) are reproduced here (Fig. 17.1). The occurrence of bi-modality in length frequency composition of some year-classes from the commercial catch had been pointed out by Felin *et al.* (1950), and as Felin (1954) remarked, this was further evidence "that pilchard caught along the Pacific coast do not constitute a single homogeneous population".

The group considered here as represented by the fish from the British Columbia fishery has been suggested by Radovich (1962) to be a genetically distinct sub-population. He called this the "far northern stock" and assumed it to have a range from off southern California to off northern California during winter, and from off central California to off British Columbia during the summer. A more northern distribution might occur in warm years and a more southern distribution during cold years. If this stock did exist, its virtual extinction "could have been the primary cause of the decline observed in the sardine fishery" (Radovich, 1962, p. 133). Virtual extinction of this stock would account for the complete collapse of the sardine fishery in the Pacific north-west. It could also explain the differences in length-on-age curves described by Marr.

Recruitment

In the first instance this refers to additions to the stock as a result of spawning. Secondly, it can refer to additions to the fishable stock as they reach a size at which they become accessible to the fishery.

In considering the first definition there are several factors which interact to determine the number of fish which result from a spawning; that is the number of fish constituting the year-class. There has, as already mentioned, been much controversy on this point, because it is on the question of whether man's influence in reducing the size of a spawning stock is a cause of small year-classes, that many questions of fishery management hinge.

Plainly, if the size of the spawning stock has no relationship to the size of the year-class produced then fishing can be encouraged without restriction. The resource can be treated as, for example, a mineral mine, except that by definition the end result will not be depletion of the resource but a continued economic benefit. Conversely, if it can be shown that small year-classes result from small, and large year-classes from large spawning stocks, then steps

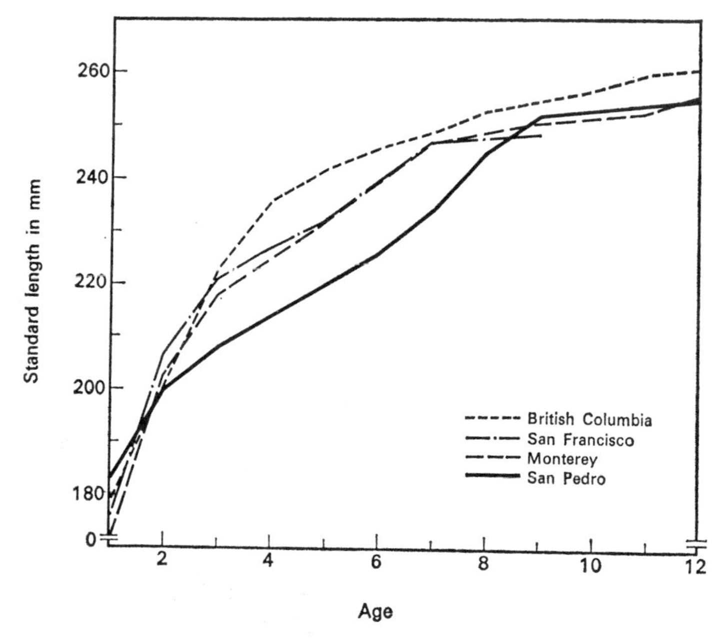

FIG. 17.1. Length-on-age curves of sardines from the Pacific coast of California and British Columbia.

should be taken to regulate the fishery. The aim of the regulations would be to adjust the catch to the economic optimum reconcilable with biological opinions regarding the maximum sustainable yield of the stock.

In the 1880's, of course, Thomas Huxley expressed the opinion that the resources of the sea were inexhaustible. Given the stage of technological advancement that had been reached at that time he was probably correct. He stated in 1883 that: "the cod fishery, the herring fishery, the pilchard fishery, the mackerel fishery and probably all the great sea fisheries, are inexhaustible: that is to say that nothing we do seriously affects the number of fish. And any attempt to regulate these fisheries seems consequently, from the nature of the case, to be useless." (Quoted in Graham, 1943.) Fifty years later when Russell (1931) wrote his classic paper on the "over-fishing" problem the fishing environment was very different and the clear mathematical statement of his case caused many people to reconsider the stocks of the sea in a different light.

Briefly summarized, Russell's thesis was that any fishable stock consisted of fish of catchable size (l) and those of less than catchable size. Due to natural processes fish of less than l grow to become part of the catchable stock and this addition to the catchable stock he called A; older fish already in the catchable stock increase in weight and this addition he called G; and, last of the natural processes, fish die and reduce the weight and numbers of the catchable stock, this process being designated by M. The numbers and weight of the catchable stock are, of course, also reduced by fishing mortality, designated by C. If the stock is considered over the period of a complete season then the total weight at the beginning of the first season (S_1) compares with the total weight at the beginning of the second season (S_2) as follows:

$$S_2 = S_1 (A + G) - (C + M)$$

S_2 will, therefore, be greater than, smaller than or equal to S_1 depending on whether $(A + G)$ has been greater than, smaller than or equal to $(C + M)$.

Russell demonstrated that reduction of the number of spawners would lead to a reduction in the stock. He stated that

> an indefinite increase in the rate [of capture] would lead to a virtual extermination of the stock, first by reducing to indefinitely low numbers the existent stock and, through destruction of the spawners, reducing A in course of time to nearly zero. This is, of course, a limiting case, never actually met with so far as fish are concerned, but it brings out the possible danger to stock through undue destruction of spawners brought about by very intensive fishing.

There are, however, workers who still believe in the inexhaustibility of pelagic fish stocks. For example, an economist, Scott (1964), suggested that "female herring produce so many eggs that it is almost unthinkable that one year's yield should have any relationship to the number of females caught the year before". Moreover, he concluded that "the concept of a *biologically* optimum catch is meaningless, since biology cannot indicate any particular size of stock or yield as an objective goal". The conclusion is logical given this initial premise of the inexhaustibility of pelagic stocks of fish, but from the viewpoint of a commercial fishery it would seem that the premise is false. It suggests that there can be no interrelationship between biological and economic schools of thought and denies the efficacy of such proven cases of control as illustrated by the Pacific halibut (for some details of this case see Thompson, 1929 and Crutchfield and Zellner, 1963) even though he had stated earlier that halibut was one of the stocks that might be overfished, having admitted that demersal fish possibly, but pelagic fish never, may be in danger of overfishing. It seems that Scott's main point is that it is most likely that a fishery will reach a point of economic overfishing before a stock will be destroyed. This is, of course, correct in most cases but the destruction of a stock is not a criterion for judging whether biological overfishing has occurred. Furthermore, to suggest that the biologist cannot indicate the optimum stock size is like saying that the agricultural scientist cannot advise the farmer on the most suitable way of obtaining a sustained yield from his land, and incidentally if this were true it would vitiate the work of biologists in estimating optimum yields, maximum sustainable yields and so on from given population sizes, all of which are aimed at rational exploitation of a fishery.

It must, however, be admitted that there have been several papers written specifically relating to the Pacific sardine, notably by Marr (Clark and Marr, 1955; Marr, 1960a) which support the idea that the fecundity of this pelagic fish is such that no relation exists between the size of the spawning stock and the resultant year-class size. Thus Marr (1960a) restated his thesis as put forward in the paper with Clark and wrote: "while changes in population size were obviously attributable to changes in year-class size, changes in spawning stock size had no demonstrable effect on year-class size over the range of stock sizes observed, and he concluded that the causes of variations in year-class size must be sought in the environment."

Other workers, while not suggesting that there is a constant correlation between spawning stock size and resultant year-class size have maintained that some relationship between the two exists, at least over a certain range of values. Clark (Clark and Marr, 1955) concluded from her studies of stock sizes, year-class sizes and spawning stock sizes, that there was a critical spawning stock size below which the size of the year-class produced depends on the size of the spawning stock producing it. Above the critical spawning stock size she concluded that environmental factors played the dominant role in determining year-class size until another upper critical value was reached, above which there may again be a relationship between the two factors. Clark summarized her interpretation of the data as suggesting that there is:

> a stock size at some point between four and eight billion [which] might correspond to the theoretical value (Ricker, 1954) at which the resulting year-class is the least dependent on stock size, thus representing the optimum stock size for greatest recruitment. Below this range (four to eight billion) year-class size may be in part dependent on stock size because too few eggs are produced. Above this range year-class size may again be partly dependent on stock size.

Marr presented an analysis which, as already indicated, stated that there was "no apparent relationship" and that the correlations Clark put forward in support of her thesis were not statistically significant. However, using the same data Radovich (1962) stated, "I have demonstrated how a relationship between spawning stock size and year-class size can exist using data with which Marr concluded 'variations in survival rate are so large that they obscure any theoretical relationship which may exist between stock size and year-class size' " (Clark and Marr, 1955).

Radovich's relationship between spawning stock and year-class size was mediated through environmental conditions. He proposed that a positive correlation could be assumed between increasingly favourable environmental conditions, and the size of the population it was possible to maintain. Radovich's figurative representation of this hypothesis is reproduced in Fig. 17.2. In this, curve A represents the relationship between year-class size and

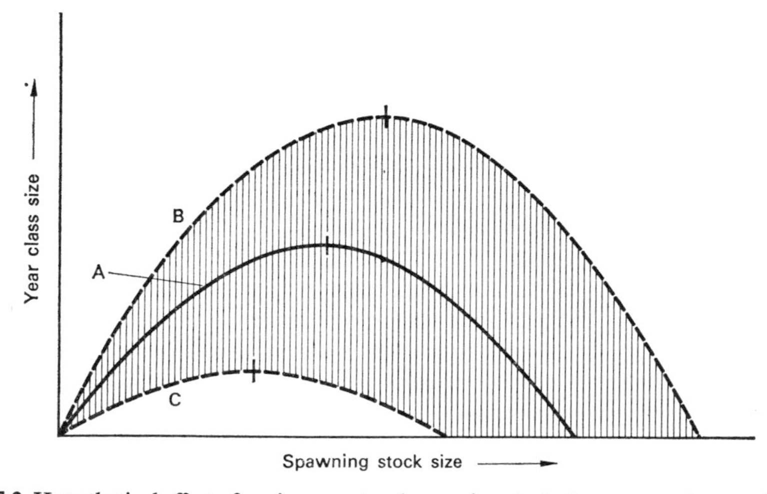

FIG. 17.2. Hypothetical effect of environment and spawning stock size on year-class production.

spawning stock size under average environmental conditions held constant. Under the most favourable environmental conditions possible the relationship is indicated by curve B, while curve C shows the relationship under the worst possible conditions.

Curve B shows that under these favourable conditions at all spawning stock sizes larger year-classes are produced than under average conditions; and the converse applies for the worst environmental conditions. The shaded area covers the whole range of environmental conditions and the consequent relationship between the two factors being considered. There are four main points of interest indicated by the diagram. These are:

1. That from a given spawning stock size a wide range of year-class sizes can be expected depending on the prevailing environmental conditions.
2. That larger spawning stock sizes beyond the maximum point of the parabolas give decreasing survival.
3. That a small increase in spawning stock size at low population levels has a much more incremental effect on the year-class size than a similar increase near the optimum.
4. From the fishery angle the converse of (3) should be stated, namely that a similar quantity of fish removed from (a) the spawning stock which is at the size represented by its optimum, for any given environmental conditions, has a much less detrimental effect on the year-class production than if it were removed from (b) the spawning stock which is somewhat below its optimum size.

The first two points dealt with above are readily substantiated from the statistics collected from the Pacific sardine fishery. Thus, similar spawning stock sizes in 1937, 1938, 1939 and 1944 gave widely differing year-class sizes. Relevant figures (from Clark and Marr, 1955) are shown in Table 31. It should be made clear that the spawning stock size is that estimated for the year given in the first column. The year-class size refers to the size of the population the spawning stock gave rise to, estimated when the fish entered the fishery at 2 years old.

TABLE 31. ESTIMATED YEAR-CLASS SIZES AND SPAWNING STOCK SIZES FOR 1937, 1938, 1939 AND 1944

Year	Spawning stock size (billions of fish)[a]	Year-class size (billions of fish)[a]
1937	4·4	3·8
1938	4·8	4·9
1939	4·9	7·2
1944	4·6	0·9

[a] Billion here and in the following pages is used in the American and French sense of a thousand millions as opposed to the British interpretation of a million millions.

The second point, that larger spawning stock sizes tend to give rise to smaller year-classes, can be illustrated by considering the 19-year period from 1932 through 1950. From the estimates of Clark and Marr it can be determined that the average spawning stock size for the period was 5·3 billion fish. During 9 years of the period the spawning stock size was above the average and for six of these above average years the resultant year-class was below the average size for the period, of 2·53 billion fish. The relevant figures are shown in Table 32.

TABLE 32. SPAWNING STOCK SIZES ABOVE AVERAGE FROM 1932 THROUGH 1943 AND THEIR RESULTANT YEAR-CLASSES

Year	Spawning stock size (billions of fish) av. 5·3	Year-class size (billions of fish) av. 2·53
1932	5·9	4·9
1933	6·7	2·3
1934	8·7	1·5
1935	8·7	2·5
1936	7·2	4·3
1940	6·0	3·0
1941	8·0	1·9
1942	7·6	2·3
1943	6·0	1·4

The 3 years in which above average spawning stock sizes gave rise to above-average year-class sizes are seen to be 1932, 1936 and 1940.

It is more difficult to show that point 3 can be factually supported from the statistics, but this can possibly be done by considering the ratio between year-class size and spawning stock size. Radovich, in the paper already referred to, called this ratio the survival ratio. If the survival ratios are considered for the 19 years from 1932 through 1950, then in the years for which the spawning stock size is less than for the previous year the survival ratio would be expected to rise when compared with its value during the previous year, and vice

TABLE 33. SPAWNING STOCK SIZES, YEAR-CLASS SIZES (figures from Clark and Marr, 1955) AND SURVIVAL RATIOS (YEAR-CLASS SIZE/SPAWNING STOCK SIZE) (figures from Radovich, 1962), 1932 THROUGH 1950

Year	Spawning stock size S (billions of fish)	Year-class size Y (billions of fish)	Survival ratio Y/S
1932	5·9	4·9	0·8305
1933	6·7	2·3	0·3433[a]
1934	8·7	1·5	0·1724[a]
1935	8·7	2·5	0·2874
1936	7·2	4·3	0·5972[a]
1937	4·4	3·8	0·8636[a]
1938	4·8	4·9	1·0208
1939	4·9	7·2	1·4694
1940	6·0	3·0	0·5000[a]
1941	8·0	1·9	0·2375[a]
1942	7·6	2·3	0·3026[a]
1943	6·0	1·4	0·2333
1944	4·6	0·9	0·1957
1945	2·9	0·8	0·2759[a]
1946	1·3	2·0	1·5385[a]
1947	0·9	2·6	2·8889[a]
1948	1·7	1·8	1·0588[a]
1949	3·0	0·05	0·0167[a]
1950	2·6	0·01	0·0038

[a] Indicates the years for which the relationship between spawning stock size and survival ratio, described above, holds.

versa. It is not possible to take variations in the favourability of the environment into account but even so this holds good for 12 of the 18 years for which the test is applicable as Table 33 shows.

A further interpretation of the statistics has been presented by MacGregor (1964) who showed that there was "a high degree of inverse correlation between spawning stock size and year-class size (based on indices of total fish caught)". He went on to state that: "the data indicate that spawning stock size influences year-class size and that this is not merely a secondary result of year-classes occurring in cycles which generate similar cycles of spawning populations several years later."

It seems that the data are amenable to interpretation in two opposing ways. Different workers have approached this difficult problem from different angles but it can be seen that the majority believes that there is some direct connection between spawning stock size and year-class size. Some workers think that this is likely only at very low population sizes. For example, Marr (Clark and Marr, 1955) suggested that, "year-class size is a function of spawning stock size only at or below the smallest observed stock size". Other workers, as has been illustrated, believe that the relationship is much more permanent than this. Evidence seems to be weighted in favour of this latter view and an inclination to this belief enables the efforts of the conservationists to limit the seasonal catch, described in the following pages, to be more easily understood.

It can now be seen that a great deal of work concerning the biology of the Pacific sardine has been carried out in California, and that every effort has been made to relate this to the commercial fishery. In the pages that follow brief details of the fishery will be given and then an attempt will be made to see perhaps some ways in which the management of the fishery may have been mistaken. It was indicated earlier that with such a biologically well-documented industry it is difficult to imagine how serious mistakes can have been made. However, it will be shown how some workers, at least, feel that occasional mistaken decisions in management may have led cumulatively to the downfall of the industry in the 1940s and, thereby, to the state of the industry at the present time.

CHAPTER 18

THE FISHERY AND THE CAUSES OF ITS DECLINE

The Fishery

The Pacific sardine is a very active fish and, like its close relative in the English Channel is best captured at night as otherwise it avoids the nets with considerable skill. It is captured most successfully when the moon is not bright and most fishing taking place during phases of the moon known as "darks", when the moon is not full. This enables any phosphorescence caused by the movement of the fish in the water to be clearly seen, thus aiding in location of the shoals (Clark, F. N., personal communication, 1966).

The seine-type of net is invariably used for the commercial catches. In the past, some fishermen used the type known as a lampara net and others used a net known as the Japanese round-haul net. These differences reflected the origin of the fishermen as the lampara was introduced to California by the Italians and the round-haul net by the Japanese. Both types, however, were of the same basic principle but differed in the details of construction and of the relative sizes of the various parts. For example, the lampara was generally between 150 and 200 fathoms long, the Japanese round-haul net between 200 and 250 fathoms long; the relative depths were 15–20 fathoms for the lampara and 25–30 for the round-haul net (Higgins and Holmes, 1921). The Japanese introduced a modification to their nets which enabled the bunt of the net to be lifted during hauling thereby developing a purse-seine action. Later, when power winches became available, the purse-seine itself was used throughout the fishery, this transition taking place around 1929–30.

A haul is made in the following way. One end of the net is cast off from the main vessel and this is either attached to a buoy or to a skiff. The latter, or if the net is attached to a buoy, the main vessel, encircles the shoal and both ends of the net are then taken aboard the main vessel ready for hauling. In vessels where the power winches are still in use the ropes are drawn in by the winch until the net is near enough to the main vessel for the fish to be extracted. The crew formerly assisted this operation by hauling the wings of the net over the side of the boat (Higgins and Holmes, 1921), but now the net is completely hauled in by the power winch. The fish are unloaded from the net by dip nets operated from power winches (Clark, F. N., personal communication, 1966). When all the fish are in the hold the remainder of the net is hauled in and the crew also have to partake in this operation.

The whole of the hauling of the net can now be done automatically since the invention of the power block by Mario Puretic in 1953. Puretic, himself an experienced sardine fisherman from San Pedro, invented a block through which the whole net would pass. To quote Schmidt (1959), Puretic "conceived the basic idea of passing the entire purse-seine net through an elevated free-swinging, self-powered V-sheave, so constructed that gravity would wedge the net into the sheave, giving it the necessary traction to pull the net out of the water".

Unfortunately this invention came after the fall of the sardine industry and very few sardine boats have changed to this method.

Statistical Survey of the Catches of Pacific Sardine

Catch figures for the Pacific sardine are complete from the 1916–17 season, although these figures do not include the amount caught for bait nor that caught for sport fishing. Figures of the landings are shown in Table 26 and they have been converted from the American measure of the short ton to the metric ton which is now used internationally by the Food and Agricultural Organization of the United Nations.

It is usually stated that it was the stimulus of the First World War which caused the California sardine industry to develop. As Table 26 shows, a very great increase in landings of the sardine in California occurred between the 1916–17 season and the 1918–19 season which tailed off slightly at the end of the war during the 1919–20 season. The reasons for this pattern were fourfold. First, the import of sardines and sardine-type products was halted during 1917 and 1918, so that there was a large unfulfilled home demand in the United States. Secondly, this was accompanied by an internal campaign in America which stressed the importance of fish as a protein substitute for meat, as meat was increasingly difficult to obtain. Thirdly, any imports of this type of product that were still able to be purchased by the United States had increased in price, due to increased production difficulties and the European war demand, so that these products were no longer so acceptable in the United States. Finally, because European countries were finding it increasingly difficult to manufacture canned fish foods, a very large overseas market was opened up for American canners.

The way in which the canners responded to fulfil this new stimulus of demand is shown by the following figures. In 1916 eleven canneries which had previously been packing tuna were converted for sardine canning. The following year (1917) eleven new canneries were

TABLE 34. CALIFORNIA SARDINE CASE PACK IN POUND-OVAL CANS AND EQUIVALENTS IN OTHER SIZED CANS

Year	Cases	Year	Cases
1916	106,745	1932	1,001,845
1917	408,666	1933	1,422,860
1918	747,737	1934	2,051,726
1919	946,069	1935	2,532,762
1920	951,793	1936	2,820,549
1921	366,191	1937	2,983,204
1922	697,643	1938	2,464,263
1923	1,088,564	1939	3,326,162
1924	1,336,554	1940	3,188,089
1925	1,687,780	1941	5,395,284
1926	2,082,631	1942	3,866,785
1927	2,403,272	1943	3,400,106
1928	2,496,966	1944	3,734,660
1929	3,489,910	1945	3,884,424
1930	3,069,524	1946	3,102,232
1931	1,794,991	1947	1,814,050

From Statistical Reports of California Department of Fish and Game (supplied by Dr. Clark).

built, followed by another ten in 1918, five in 1919 and two in 1920. The catch figures as given in Table 25 indicate the large increase in total catch from California from 24,970 tons in 1916–17 to 60,800 tons in the 1919–20 season. It appears from the statistics that the 1918–19 season was a peak year in the period from 1916–17 to 1921–22 and the reason for this is not clear. However, it is likely that the artificial boost given to the industry by the war had ended in 1919, and as canneries were reopened in Europe the competition increased, giving a temporary setback to the industry. Figures for canning production in California during the war period confirm this pattern (Table 34) but show a time lag so that the peak for the period up to the end of 1922 came in 1920. In 1921 there was a marked fall in output of canned sardines but this was followed by large annual increases up to the end of 1930.

To return to consideration of the landings, for the next 14 years from 1922–3 through 1936–7 apart from minor reductions from one season to the next during four seasons, there was a remarkable upward trend. Therefore, if it is allowed that the landings had been artificially high for the stage of development of the industry during the First World War, then it can be stated that for the twenty-one seasons from 1916 through 1936 the industry showed an extremely rapid development taking quantity landed as a criterion. Thus, if the landings for the seasons 1918–19 through 1922–3 are totalled and compared with the total for 1923–4 through 1927–8, the total for the second 5 year period is more than two and a half times that for the first. The respective totals being 256,390 tons and 665,565 tons.

Using the criterion of landings again, in 1937–8 the industry started on its decline. However, the 1936–7 landings were exceptionally high and the landing figures were actually maintained at a very high level from 1934–5 through 1945–6. The canning production figures (see Table 34) show that there was an increased output for human consumption during the years 1939–46 inclusive. Increased demand from Europe for this type of food during the beginning of the Second World War, followed by America's entry into the war meant that all available food resources were put into products for human consumption wherever possible. It was in the following year of 1946–7 that the catch fell to a lower level than it had reached for 14 years (211,875 tons), and the following year the landings were nearly halved again and fell to 110,050 tons. The recovery of the landings which occurred in the 1949–50 and 1950–1 seasons, probably due to the good survival of larvae (Radovich, 1962) from the 1946, 1947 and 1948 spawnings, was short lived. The landing figures from 1951–2 to 1963–4 have fluctuated very widely, which is typical of a fishery based on only one or two year-classes with no older fish as a reserve resource to bolster unsuccessful spawnings.

At this stage it is constructive to ask some questions prompted by the foregoing survey of the landing statistics of the industry. These questions are:

1. What stimulus or stimuli caused the rapid development of the industry between the mid-1920's and the mid-1930's?
2. Was the decline and fall of the industry due to overfishing?
3. If the downfall of the industry was due to overfishing, how did this occur when so much research was being, and had been done on the resource? If it was not due to overfishing what other possible reasons for the decline can be suggested?

Before attempting to answer these questions, however, it is interesting to look briefly at the history of the industry in British Columbia. Here the fish, *Sardinops caerulea*, is known as the pilchard and the fishery for it started in 1917–18. The maximum catch for the period from the initiation of the fishery there until 1921–2, was in 1920–1 when just over 3500 tons were landed. Although this fishery was probably also started as a war effort it continued until the end of the 1940s. The landings were very low until the 1926–7 season when they reached nearly 44,000 tons. From that season the average for the 20 years until the 1945–6 season was just over 51,000 tons. Although this in itself represents quite a sizeable fishery, the quantities landed were insignificant when compared with the California figures (figures from Marr, 1960a).

To return to the questions posed above, each will be considered in turn.

1. It has already been pointed out that the initial impetus to the development of the industry was provided by the First World War, and that after the war had ended the industry lost some ground, probably due to overseas producers re-entering the market for canned sardines. After that, in the early 1920s, canning gained ground again and production began to rise. It was always the policy of the workers connected with the industry via the biological research stations of the California Fish and Game Department, that fish should be used as food for direct human consumption whenever possible. Any mass reduction of food resources for preparation of meal for animal food or for oil not for human consumption was considered an abuse of a resource. However, from all canning plants there is inevitably a certain amount of offal, as the head and viscera are necessarily discarded before a fish can be canned. Further, some fish may be broken, or rather thin and wasted, perhaps as a result of breeding, and would not make a satisfactory canned product. In order that such offal should not be wasted, in 1920 reduction of the material was made legal, but only so that such material should not be wasted and no further reduction was allowed. The quantity of this sort of material probably amounted to from 40 to 50 per cent of the catch. However, the terms of the 1920 permits were apparently rather ambiguous, and in the following year the wording of the permits was couched in more quantitative terms. The quantities involved were based on the intake of the canneries concerned so that at San Pedro 25 per cent of the intake of a cannery could be reduced and at Monterey 10 per cent, "the difference being deemed necessary because of the difference in fishing conditions near the two ports" (Schaefer *et al.*, 1951). This differential was, however, removed the following year (1922) as the Commission of Fish and Game in California considered that the Monterey canners were placed at an unfair disadvantage by it.

The quantitative basis was changed again in 1925 when it was permitted for canners to reduce an amount of fish equivalent to 25 per cent of the capacity of the canning machinery of the factory. It is unfortunate that the basis of the reduction laws had changed by this time. In an attempt to avoid ambiguity the original intention of allowing reduction to avoid waste had by 1925 become forgotten. The new quantitative basis had in effect made conservation more difficult by being explicitly permissive. Furthermore, it was not stated that the products of reduction from the cannery offal should be for human consumption so that these products found a ready market in the industries shortly to be enumerated. Until 1925 no concerns other than those already operating canning plant had been permitted to undertake reduction of any sort. In 1925 the regulations were broadened somewhat so that other firms could operate reduction apparatus under licence issued by the California Department of Fish and Game provided that the products could be used for human consumption.

There was, however, by this time a market for fish meal and oil to be used in products other than those which were utilizable for human consumption. The menhaden industry had a considerable history of fishing for reduction purposes and was already quite well developed as an industry by 1850 (Hatton and Smalley, 1938). In those early days menhaden oil was largely used as a substitute for linseed oils in outdoor paints. In 1865 it was further used in the curing processes for leather. With sardine oil it was widely used in the manufacture of paint, varnish, linoleum, the felt-base paper industry and the waterproof fabric industry and in the manufacture of soap. Smaller amounts were used in insecticide sprays, in the manufacture of candles, in the leather industry for tanning, in the manufacture of artificial leather and rubber and other minor uses.

Meal was also used as food stuffs for animals and poultry although initially difficulties were experienced in removing the fishy flavour. This was necessary because if too much fish meal was eaten by animals then either the flesh or the eggs had a fishy flavour. Another problem which had to be overcome in this sphere was the drying of the meal which was unstable when it contained more than about 12–15 per cent of moisture. Also excess moisture meant that the value of the meal was relatively reduced on a weight for weight basis. These two problems were overcome, however, with the advance of technology which enabled the oil to be extracted from the meal, thus removing its fishy flavour, and by the development of plant which could dry the meal without impairing its quality.

The numerous outlets for the products of reduction meant that the demand was steady and quite strong, but as yet the regulations of the California Commission of Fish and Game had not allowed the sardine resource to be used to fulfil the demands, except for the offal mentioned above.

Later regulations allowed canners to reduce a certain percentage of their intake, regardless of the condition of the fish, provided that a certain minimum amount had been canned. Thus, in 1929 it was stated that every ton of sardines should produce twenty cases of 1-lb cans or the equivalent, each case containing forty-eight 1-lb cans. Every ton of sardines suitable for canning should, in fact, produce twenty-two cases so that the regulations were allowing a fair margin to account for broken or small fish (Scofield, W. L., 1938). In spite of this adequate allowance, over the next few years the quantity of a minimum of twenty cases per ton was gradually reduced to eighteen, fifteen and then to thirteen and a half per ton so that the amount available for reduction was gradually increased. This increasing margin of fish allowable for reduction became known as "overage". It is noticeable that a change of emphasis had crept into the regulations so that now a minimum quantity had to be canned, whereas before the regulations had aimed at a minimum quantity to be reduced.

Nevertheless, the declared aims of the California Commission of Fish and Game have always been, and were at the time, conservation of a resource and, as the earlier quotation from Thompson (1926) showed, the Department of Fish and Game was well aware that any exploitation of a resource must be "rational and restrained". In spite of the change of emphasis in the regulations mentioned above, it was still considered that the utilization of a food resource such as that of the Pacific sardine should be primarily for food for human consumption. This attitude was made adequately clear in 1926 when a company was formed to operate a floating barge for reduction purposes alone and anchored it in Monterey Bay. The company concerned considered that the *Peralta*, as the vessel was called, was outside territorial limits and therefore beyond the jurisdiction of the State and beyond the regulations restricting reduction to products solely suitable for human consumption. The company asked some fishermen to deliver their catches to them and this would then be reduced aboard the vessel. However, the State authorities brought orders against the owners of the vessel and the fishermen involved. After a series of legal discussions the *Peralta* was deemed to be within territorial waters when the territorial boundaries were drawn between headlands. The owners of the *Peralta* were sued and subsequently moved the vessel outside territorial waters but were forced to close down as no fishermen would then work for them.

In the same season another floating plant known as *Lake Miraflores* started operations off San Pedro but after the *Peralta* case this, too, closed down and did not attempt to resume operations for some years. In this case, although apparently anchored beyond territorial limits, the plant was more or less forced to close down as fishermen were loth to be associated with such an anti-authoritarian concern.

In 1929 new permits were issued to the reduction concerns which were not also canning a percentage of their intake, to enable them to produce "edible products only" (Schaefer *et al.*, 1951). However, this was another slight but significant change in emphasis from the wording of the permits which the new ones were replacing. The older permits were for the production of items for human consumption. These new permits meant that the reduction firms were able to join in the competition for the fish meal market which, since the oil could now be extracted from it, was a doubly lucrative industry, there being growing demands for the oil and the animal foods.

To summarize these events it has been shown that in the decade from 1920 to 1929 it had been made legal for canners to reduce offal to prevent waste, and later they had been given a rather freer hand to determine the amount reduced, within certain limits, by the introduction of the concept of overage. In 1925 non-canning concerns were allowed permits and in 1929 these non-canning firms were allowed to enter the fish meal market. The regulations which enabled canners to reduce a certain percentage of their intake encouraged them to increase their intake so that they could take full advantage of the profitable meal and oil markets. The canning part of their business was increasingly becoming the cover whereby they could obtain permission to reduce sardines. The effects of these changes in permits can be seen in the increases in landings. Thus, in the 1920–1 season landings were 34,880 tons, in 1925–6 they had more than trebled and were 124,507 tons and in the 1928–9 season they were 230,900 tons, almost double the 1925–6 figure and seven times the 1920–1 figure.

The next factor which played an important part in the growth of the industry was, ironically, the economic depression of the early 1930s. The situation was aggravated by a floating reduction plant which started operation beyond State jurisdiction in the 1930–1 season. Although fishermen were rather slow to co-operate with the floater, as this type of plant later became known, it nevertheless reduced about 10,000 tons in its first season. The following season it trebled its intake and when its success was obvious to others it became joined by further floating plants. According to Schaefer *et al.* (1951) there were two floaters operating in the 1932–3 season, four in 1933–4 and nine in 1936–7. Details of numbers of plants with their intake are given in Table 35.

The competition from these unrestricted plants made the position of the shore-based

TABLE 35. THE INCREASE AND DECLINE IN THE NUMBER OF FLOATING REDUCTION PLANTS FROM 1930–1 TO 1939–40, AND THEIR INTAKE OF FISH

Year	Number of floating plants	Intake (metric tons)	Year	Number of floating plants	Intake (metric tons)
1930–1	1	9,940	1935–6	4	136,807
1931–2	1	28,154	1936–7	9	213,705
1932–3	2	53,324	1937–8	6	61,297
1933–4	4	61,514	1938–9	6	39,809
1934–5	3	101,623	1939–40	0	—

reducers and canning–reducers very difficult and they found it increasingly hard to sustain their position in the market. As the floaters increased their production so the shore-based plants found it necessary to apply for increases in their permits. The Legislature of the State of California did not act against the floaters and felt obliged to support the claims of the legal producers, considering their application for larger permits in a favourable manner (Scofield, 1938). They therefore issued permits to those reducers which were shore-based which allowed all plants to reduce 6800 tons (7500 short tons) of fish per season in 1932. During the 1933–4 season this was reduced slightly to 5442 tons (6000 short tons) but the permits now allowed the fish to be reduced for any purpose; even though by now it had become possible by hydrogenation for fish oil to be incorporated in products for human consumption without leaving a fishy flavour. In the 1934–5 season the amount per plant was doubled. The reasons for this increased leniency on the part of the California authorities were due to the two points already implied, namely the competition from the floating plants and the severe economic conditions prevailing at the time. The motives of the authorities here were undoubtedly philanthropic and helped towards providing employment and some local prosperity, but as Scofield (1938) wrote, that "heroic depression treatment was proving to be more of a curse than a blessing" having regard to the long-term well-being of the industry. In 1935–6 the allowances per plant were reduced quite considerably over those for the previous season to about 3920 tons (4320 short tons). This measure would probably have been beneficial if maintained, but the allowances were only reduced to this level for the one season.

It can be seen that requests for increased permits for reduction had been granted by the California State Legislature and that the reaction of the Legislature to the existence of floaters was basically incorrect, if they wished to exploit the resource rationally. It must be stressed that advice from the biological research workers connected with the division had been consistently ignored as will be shown when an answer to the second question is considered. It would probably have been preferable if the State had taken a firm line with the floaters immediately they attempted to enter the reduction business. Even if they were completely beyond all sanctional proceedings on land, which seems unlikely, they could have been effectively stopped if fishermen had been forbidden to deliver fish caught in State waters to points beyond the jurisdiction of the State. This is eventually what had to be done. The measure was inaugurated in the 1939–40 season and immediately brought the business of the floaters to a halt after which they never restarted. However, given the existence of the floater the powerful commercial interests of the sardine industry had strong moral pressures which they effectively brought to bear upon the Legislative bodies connected with the issuing of permits.

The landings themselves from the 1930–1 season through the 1935–6 season can now be seen to have taken on a pattern which neatly, if partly fortuitously, reflected the events just described. For 1930–1 through 1932–3 they were below the level reached in 1929–30. It is probable that that was due to the onset of the economic depression. The decline was, however, only temporary and was checked in the 1932–3 season which showed a marked increase over the previous season. In 1933–4 landings were over 350,000 tons and were higher than ever before and in the following year, as a result of the increasing competition and permissive regulations, reached nearly 550,000 tons. Less than one-fifth of this amount was delivered to the floaters, the remaining approximately 440,000 tons being utilized by shore-based concerns and of this about 310,000 tons was reduced to meal and oil. (Figure, from U.S. Fish and Wildlife Service and the California Department of Fish and Game,

Clark (1952).) When the floaters' contribution is added to this, a total of over 412,000 tons of sardines were reduced during the 1934–5 season. The final season for this period, 1935–6, showed a slight fall in landings but the total was still over 500,000 tons. However, the total quantity reduced was considerably lower than that for the previous season, being just under 300,000 tons, and the quantity reduced by shore-based concerns was less than half their total for the previous season, being only just over 150,000 tons. This is still a very large amount but in relation to the previous figure was a marked improvement from the point of view of the conservationists. The reduction was due to the lowering of the amount permitted per plant and without the floaters may possibly have been maintained at that level, but it was unfortunately destined to be raised again.

2. It is now possible to consider the second question to see whether the decline of the industry was due to overfishing. The reasons for the tremendous expansion of the industry have been analysed briefly, and the survey of the landing statistics showed that from the mid-1930's to the mid-1940's landings were maintained at a high level. Nevertheless, the scientists concerned with the biological resources were not happy about several factors. These were:

1. The rapid rate at which the expansion occurred.
2. The high level which the landings had reached.
3. The effect these first two factors might have on the balance of life in the sea.
4. The more or less continued increase in the permitted amounts for reduction and the fact that reduction was becoming the chief outlet for the landings.
5. The slow reactions of the Legislature to the presence of the floating reduction plants.
6. The consistent way in which the State Legislature ignored the advice given to them concerning the possible effects of the prevailing pattern of exploitation at the time.

It is very illuminating to read the comments and recommendations that the biologists made concerning the conservation measures which they considered might be effective in developing rational exploitation of the Parific sardine.

In 1924 Thompson wrote:

> The sardine is a source of food for almost all our other great fisheries, such as the albacore, barracuda, sea bass and tuna. Tampering with its abundance may result disastrously to many interests—and in the absence of any clear cut and sensitive method of detecting overfishing, the greatest caution must be used. The writer has convinced himself that unnecessary drain on the supply should be avoided until research has shown that it is possible to detect overfishing in time, and for that reason it is his belief that the use of sardines for fertilizer should be emphatically condemned, and a more conservative growth of the fishery awaited.

Thompson was a very great fishery biologist and he had foreseen, even though when he wrote the quoted words the sardine catch of California had not quite reached 180,000 tons, that if events were to continue unchecked the fishery would be heavily exploited. He also considered that the catches were increasing too rapidly as they had trebled in eight seasons. As has been seen, no notice was taken of this timely warning; the following year the catch was doubled.

Clark (1939) pointed to indications that the sardine stocks were being overfished. The most cogent indication was that the catch per unit of effort had fallen steeply over the four seasons 1934–5 through 1937–8. "In 1937–38 the average month's catch of the individual boats was but 30 per cent as large as in the 1934–35 season." Earlier than this Scofield, N. B. (1931), had drawn attention to a similar state of affairs:

Although the amount of sardines caught has been increasing each season, the catch has not increased in proportion to the fishing effort expended, and there is every indication that the waters adjacent to the fishing ports have reached the limit of their production and are already entering the first stages of depletion. The increase in the amount of sardines caught is the result of fishing farther from port with larger boats and improved fishing gear.

This latter point was again emphasized by Clark (1939) and she also showed how the average life span of the sardine had been greatly reduced from about 10 years to 4 years in 30 years of commercial fishing. She wrote that when the stock was healthy it would probably have withstood a fishery of about 300,000 tons p.a. at the maximum, but as it was not healthy in 1939 it should probably have been held below 200,000 tons to enable the stock to recover.

Scofield, W. L. (1938), gave a brief survey of the kind of warnings which had been issued by the research division of the California Fish and Game. Thus, by 1928 the research division considered that there were serious signs of overfishing and at the State Legislature in March 1929 these signs were officially pointed out and processors were asked to can the fish rather than reduce them. "By 1930," Scofield continues, "the indications of depletion were so serious that the State officials and leaders of the sardine industry were officially warned at a meeting of the Jost Assembly Interim Committee on April 16th, 1930."† In 1933 and again in 1934 warnings of overfishing were printed and distributed and catch limits proposed. These limits were opposed on the grounds that the research of the Division of Fish and Game did not furnish positive proof of serious depletion of the stocks. Scofield aptly commented that the leaders of the industry who held this view apparently did not realize that positive proof could only be supplied by the complete failure of the fishery.

Further indications of overfishing were again demonstrated in the same paper by Scofield (1938) for the three seasons from 1935–6 through 1937–8. There were 255, 320 and 380 boats fishing for sardines in the three seasons. In 1937–8 the 380 boats caught 60 per cent of the catch of the 1936–7 season by 320 boats; and even for the 1936–7 season individual boat catches were smaller on average than they had been in previous years.

Schaefer *et al.* (1951) showed how various writers between 1932 and 1939, whilst the industry was in the middle of its most expansionist phase, were issuing warnings that the catch ought to be limited to a quota of some sort. They quoted Scofield, N. B. (1932, 1934), as suggesting that the limit ought to be 200,000 tons p.a. Scofield, W. L. (1938) who thought 250,000 tons would probably be a safe annual yield, and Clark's (1939) estimate of 250,000 to 300,000 tons p.a. The authors commented: "the power to regulate the total catch, except by regulation of the fraction to be allotted to reduction, has not been advocated by the State's conservation agency for many years."

3. From these extracts by various writers it would appear that the majority of the biologists connected with research on the Pacific sardine were of the opinion that the resource was continually overfished from the late 1920s onwards until the late 1940s. Throughout this period landings were held at a high level due to increased demands for fish meal and oil but not for canned sardines. The decline in the landings, which started in the 1945–6 and 1946–7 seasons, is generally considered to have been brought about by the severe overtaxing of the resource over 20 years and particularly over the 14-year period from 1932–3 through 1945–6.

† When controversial bills are introduced into the California State Legislature, such bills are frequently referred to interim committees for further study with instructions to report back to the next legislative session. William P. Jost chaired such a committee in 1930. Hearings were held at which biologists presented their information about the sardine supply. Presumably a report was made at the next legislative session but no action taken. Many similar committees were set up in succeeding years with similar results. (I am indebted to Dr. Clark for the information contained in this footnote.)

It appears that the overfishing occurred because strong commercial concerns connected with the reduction of sardines were powerful enough to obtain their own way, regardless of the long-term effects which their exploitation might have on the resource. It also appears that these concerns were powerful enough to override the authorities directly connected with the biological research stations who were responsible for bringing about measures which should have lead to conservation and a rational exploitation of the resource. From a purely objective point of view it would seem that it must have been sheer political power which was able to obtain its own way during the early 1930s against repeated warnings of overfishing and requests for the imposition of a catch quota by stating that the results of the research of the Division of Fish and Game did not furnish positive proof of depletion, as mentioned earlier.

It can, therefore, now be understood why the resource of the Pacific sardine was overfished. The research was done and the results made public and even actively drawn to the attention of the people in a position to regulate the fishery, but these people who were really concerned about the future of the resource were not in the administrative positions to be able to obtain the regulatory action they required. They were not unaware of the difficulties involved in bringing about regulation. It was well realized that the administration and enforcement of, for example, a seasonal quota for the sardine catch was a complicated task.

Clark (1939) stated that the limit she recommended (of 200,000 tons to re-establish the fishery) would be difficult to impose, but she suggested that, either the reduction plants could arrange among themselves how the catch should be allocated which was admittedly "fraught with difficulties", or permits for the reduction of sardines to fish meal and oil could be rescinded.

Throughout most of the history of the industry there have been closed seasons for catching sardines for purposes other than use for bait, salting, smoking, curing, drying or for packing in quarter- and half-pound square cans. These closed seasons have altered from time to time and have varied with the locality, but their effectiveness to limit the catch has been minimal. This is because extra effort is usually expended during the open season. Their effectiveness in preventing the capture of too many spent fish has possibly been greater as the closed season has always been during the breeding season.

Other possible methods of regulation of the fishery have been suggested but the usual ones such as regulating mesh size and having closed areas do not work. The former because sardines are caught in purse-seines which are selective only at the extreme lower limit. The latter because sardines are migratory fish.

In the late 1940's and early 1950's the efforts of research were turned to re-establishment of a strong stock and a rebuilding of a certain reserve of year-class strength in an attempt to even out the sort of fluctuations in landings which occur when fisheries are based on only one or two year-classes. These fluctuations occur because, inevitably, there are relatively more successful and relatively less successful year-classes and the fortunes of the fishery depend on these unless there is the reserve of somewhat older fish.

Other reasons for removing the burden of the fishery from the 2- and 3-year-olds are based on attempts to utilize the resource in the most economical way. For example, the sardine maintains a fairly rapid growth in the first 3 years of its life, and removing fish of 2 and 3 years old means that the maximum growth per fish for the minimum amount of food consumed is not being attained. A fishery based on the 4-year-olds would ensure that the maximum growth had been attained by the fish in the population.

A further reason why it is advantageous to remove the fish of more than three years old is that many more younger fish, of course, are needed to constitute a ton as Table 36 shows.

TABLE 36. NUMBER OF SARDINES PER TON AT AGES 1, 2 AND 3 YEARS
(Data from Clark, 1952)

Age	Average length (in.)	Number per ton
1	7·7	14,000–16,000
2	9·1	8,000–10,000
3	10·0	6,000–7,000

As selective methods of conservation are not available for the Pacific sardine except at the lower limits, it would be necessary to regulate the mesh size of purse-seines used in the fishery to ensure the escape of fish up to 3 or 4 years old. Correlated with this there could be a seasonal quota which made certain that enough fish were likely to remain uncaught in the several yearly year-classes so that a backlog of breeding fish could be established. Clark (1952) emphasized the importance of this once again in the following way.

> To bring about the desired increase in the numbers of older and larger sardines in the population will require a sacrifice on the part of the sardine industry. Some means will have to be devised to hold the catch at a level low enough to permit a longer life expectancy for each year-class than now occurs. . . . A fixed total tonnage, commonly termed a seasonal bag limit, would hold the total catch to the desired level but methods of allotment are complex and without allotments the industry might be thrown into chaos. The final decision about satisfactory management methods should rest with the industry. It has the most to gain if a constructive management program can be developed and the most to lose if no solution is reached.

The majority of the evidence seems to indicate that the Pacific sardine is a resource which has been abused by overfishing. This occurred, as has been seen, in spite of constant warnings from biologists. The fecundity of a pelagic species such as *Sardinops caerulea* is without question and has never been questioned, and that the species will become extinct is against all reasonable opinion. However, from the point of view of maintenance of the resource so that a continued economic benefit can be obtained the Pacific sardine has been mishandled. It was pointed out in Chapter 1 that man is inept at conservation of even what appear to be his most flourishing resources and that he continues on his destructive course in spite of numerous signs of danger of over-exploitation. One of the most useful lessons to be learnt from the comprehensive research programme that has taken place in California for the sardine is that the industry would be healthier now if it had heeded the early advice tendered to it. If this can be applied to resources which can still be preserved, such as the whale stocks and the South African pilchard stocks, then the research workers will have made a valuable contribution to the well-being of mankind.

It has already been mentioned that Marr (1960a) has proposed that the decline of the sardine industry was not due to overfishing as he considered the fecundity of the species great enough to make year-class size independent of spawning stock size. He suggested that the size of a year-class was dependent on environmental conditions and that a series of changes had occurred in the environment of the spawning grounds. These changes were unfavourable to the sardine and included a lowering of the average temperature, and a raising of the salinity during a period of above average north-westerly winds as measured at the Pier of the Scripps Institution of Oceanography. They have resulted in a lowering of the success of spawning since about 1943. Marr concluded from his environmental studies that, "a major change in the oceanographic regime has occurred off southern California, and that this change has coincided in time with a change in sardine year-class size". He further proposed that these changes had acted in favour of the anchovy which is able to compete more successfully with the sardine under the lower temperature conditions having a lower temperature threshold for the survival of the larvae.

Environmental studies such as that undertaken by Marr are complex and the results are very valuable and it may well be that they will, in due time, be able to provide the basis of forecasts for various fisheries. It is also apparent that the environment plays an extremely important part in determining the size of any year-class. It seems likely that the environmental changes discussed by Marr have been powerful influences in the decline of the fishery but it is also felt that the fishery would not have become so reduced had fishing not been so exhaustive in past years. This opinion is supported by Murphy (1966) who, in a later paper, examined again the relationship between spawning stock size and reproductive efficiency. He stated "that it is improbable that the population would have declined in the absence of fishing, whereas the fishing rates applied to the population lowered reproduction to an extent that a decline was inevitable."

It was mentioned above that one of the problems of the sardine industry in recent years is that the landings have had to rely on only two or three year-classes for their success or failure, that the fishery has been exhausting the new year-classes before they have been able to fulfil the most incremental parts of their growth curves and that as a consequence a reserve of year-classes has not been allowed to accrue. At this point, therefore, it seems reasonable to suggest that the prolonged heavy fishing which occurred from about 1928–9 through 1945–6, combined with the unfavourable environmental changes, have resulted in the failure of the sardine industry in California. However, if the industry had heeded the warnings of the biological workers there would have been much more latent strength in the older year-classes for the resource to draw upon to survive the poorer years which would perhaps in any case have been brought about by the environmental changes.

CHAPTER 19

THE PRESENT STATE OF THE INDUSTRY AND POSSIBILITIES FOR THE FUTURE

IT IS necessary at this stage to look briefly at the ways in which the landings of sardines have been utilized in California since the Second World War. From the previous pages it will have become apparent that initially the main outlet for the raw material was to the canning industry and that later as the importance of fish meal increased, so the percentage of the catch reduced gradually increased. Details of the actual canning and reduction processes will not be dealt with here, but the canning process for the Cornish pilchard has already been described and is broadly similar to that used in California. No two factories will be exactly alike in the details and the layout of the plant but the basic principles involved are identical and do not warrant separate descriptions for each country.

From the graph, Fig. 19.1, it can be seen that canning remained the major outlet for sardine products in California from the end of the war until 1964. Since 1963–4 there has been very little sardine canning, most of the catch being sold fresh and frozen for bait (Clark, F. N., personal communication, 1966). As the graph indicates, the output of oil and meal from the industry as well as the output for the canned product all follow a similar

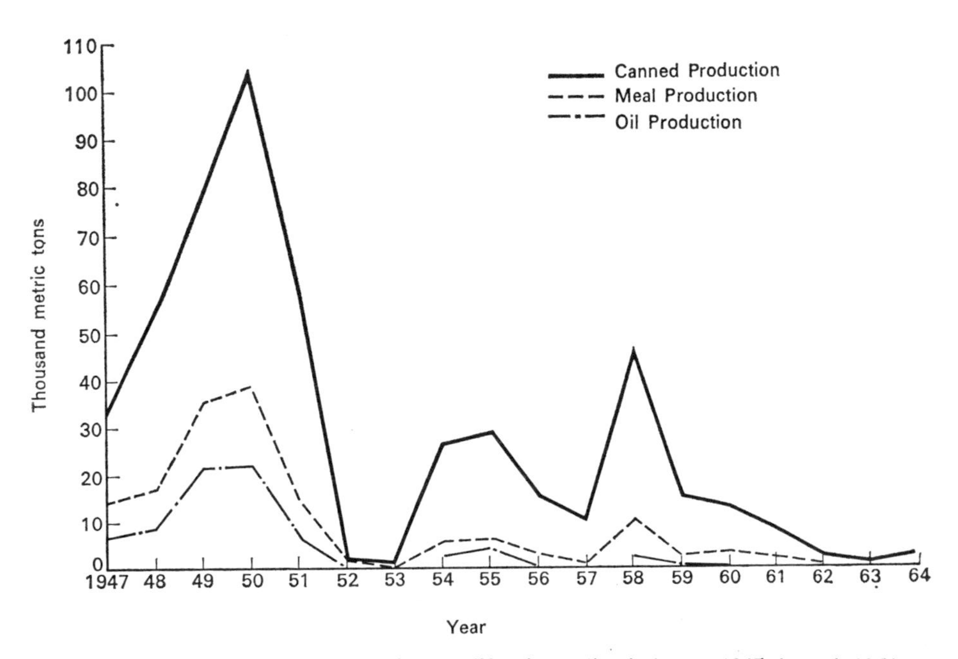

FIG. 19.1. Product output of the California sardine industry, 1947 through 1964.

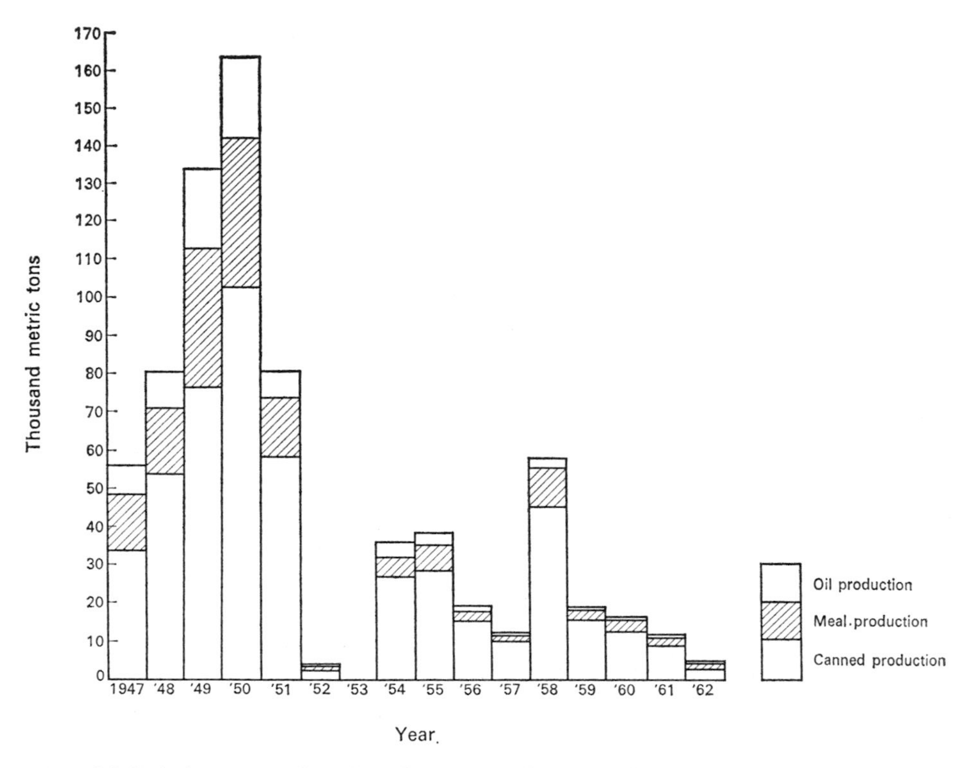

FIG. 19.2. Relative output of products from the California sardine industry, 1947 through 1962.

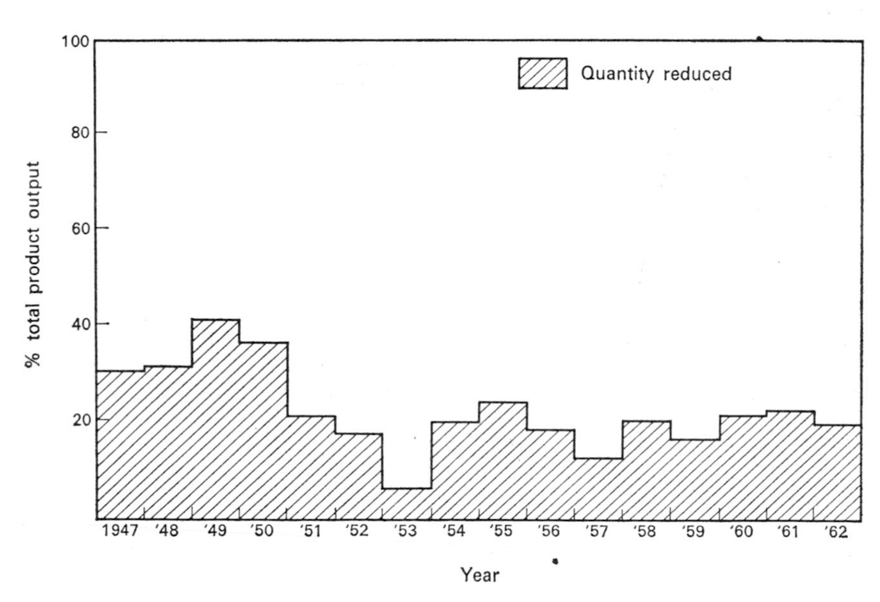

FIG. 19.3. Percentage of total production forming products of reduction from California sardine, 1947 through 1962.

pattern. This pattern has been completely dictated by the availability of the Pacific sardine. This can be ascertained by comparing the figures of the landings since 1947 with the rise and fall in output of all three products. All four factors (i.e. landings and production figures) are directly related.

Figure 19.2 shows the relative output of each product. Production figures for 1952 and 1953 were so low that they can only be roughly indicated in 1952 and do not warrant being shown for 1953. After that there was a considerable resurgence in 1954 and 1955 and again in 1959, but as mentioned earlier the landings of the Pacific sardine now fluctuate widely and as a consequence the production figures show very wide fluctuations also.

Figure 19.3 has been constructed to indicate that the products of reduction since the Second World War have been kept within very reasonable limits. When this diagram is studied it can be seen that for only 4 years in the 16-year period from 1947 to 1962 did the percentage of the production output that was reduced rise above 30 per cent. These years were from 1947 to 1950 inclusive.

It is not easy to say what the future of the industry will be, but if it remains viable it seems that the main outlet will be the canned product, the products of reduction being merged with those of other fisheries. The graph, Fig. 19.1, shows that this pattern is already emerging, production of oil and meal from 1960 onwards being barely recordable.

As far as the catching side is concerned it has already been mentioned that considerable fluctuations have occurred since 1945–6 and since the 1958–9 season landings have declined rapidly. It is possible that the decline in the size of the population of the Pacific sardine, whatever factors may have caused this, has been accompanied by an increase in abundance of the anchovy (*Engraulis mordax mordax*) in the area. In the early 1950's, when the decline in abundance of the Pacific sardine began to make itself severely felt, anchovies were sought as a suitable substitute. Landings of the two were inversely dependent on one another, at least in California, for several seasons. Miller (1956) makes the following statement on the situation which prevailed at that time: "In 1949 and 1950 sardines temporarily 'returned' to Californian waters; as a result there was lessened interest in anchovies and the catch lessened. However, when the sardines declined rapidly in 1951, anchovies were again desired and permanent markets for anchovy products were sought, for it now seemed that the sardine population was indeed at a very low level of abundance." The reader is referred to Murphy (1966) for further interesting observations on the anchovy population of the present time.

The figures for landings of anchovy and sardine in California for the 1959–60 through the 1964–5 seasons are compared in Table 37.

It can be seen that the landings do exhibit a tendency to vary inversely with one another but that this is not invariable. It can also be seen that the quantities involved are not very great even though in 1960–1 they were considerable. The small size of the landings is probably a disappointment to those who looked to the anchovy to fill the gap left after the sardines were no longer found along the coast in their former great numbers. However, research has started on the resource and it may be that the full potential is not yet being fully realized in spite of the use of aerial spotting for detecting shoals. Murphy (1965) suggested that with reference to exploitation of species occupying similar ecological areas, "judicial utilization of all ecologically similar species within a trophic level offers the only hope for sustained yields".

The future prospects for the sardine industry itself remain very poor. In the pelagic fish surveys reported in *Commercial Fisheries Reviews* there have not recently been signs that

TABLE 37. COMPARISON OF LANDINGS IN CALIFORNIA OF ANCHOVY AND SARDINE, 1959–60 THROUGH 1964–5

Year	Anchovy (metric tons)	Sardine (metric tons)
1959–60	3,250	33,700
1960–1	22,700	26,100
1961–2	3,040	19,600
1962–3	1,250	7,000
1963–4	2,100	3,200
1964–5	2,250	5,450

Figures for anchovy landings, except 1963–4, from the California marine fish catch for 1959–62, 1964. The quantities for the 1963–4 catch of anchovy and for the sardine landings from the F.A.O. *Yearbooks of Fishery Statistics*.

any improvement in landings can be expected over the next few years. Fishermen and biologists have been looking for signs of a strong year-class, which would boost the fishery, but none has been forthcoming.

It seems likely, therefore, that the fishery will die out except for the landings of the older fishermen who will probably continue to land sardines when they are available, making a modest living, in a similar way to those in Cornwall. Also those fishermen who catch mackerel and anchovy (this meant seventy to eighty boats in 1966) will probably also continue to take sardines if and when the fish become available. In a few years it is likely that the gear used in both the catching and processing sides of the industry will be extinct, having been replaced by that necessary for other fish. This, in fact, will possibly mean that if a strong year-class of sardines should survive, then this lack of gear may give it a chance to build the population up once more to fishable levels. If the fishery revives somewhat before the gear has become extinct then the population will again be fished beyond the reproductive capacity of the stock. This would be inevitable as the fishing that occurred would be an attempt to obtain a return from the investment in the gear.

The story of the sardine industry of California has been one of very great interest to biologists, economists and those others who may be able to view it from a rather detached position. To others, those more closely involved in it, the story has been rather tragic and in the years from 1945 onwards probably involved many in considerable financial losses. However, as mentioned earlier, if the research work carried out in connection with the industry can be taken as an example to other fisheries at present flourishing, to view their resource with respect, then the fishery will have been very valuable to mankind.

PART IV

THE PILCHARD INDUSTRY OF SOUTH AND SOUTH WEST AFRICA

CHAPTER 20

HISTORICAL BACKGROUND AND BIOLOGICAL RESOURCE

THE pilchard industry of South Africa and that of South West Africa are often considered together as they both exploit the same species, *Sardinops ocellata*. Research suggests, however, that the stocks are geographically distinct, there being little evidence of much interconnection in either a northerly or southerly direction. (See, however, the section on migration.) The industry in South Africa did not start until 1943, although for many years fishermen had realized that there were large shoals of sardine-type fish not far from their home ports. Individual fishermen before the Second World War were generally poor, average earnings being about R100 per year (Irvin and Johnson, 1963). (Although not strictly comparable with pre-war figures parity for one rand was 1·40 American dollars in 1966 and about 11*s*. 6*d*. sterling in 1968.) One survey taken in 1936, quoted in Stoops (1953), showed that in one area of the west coast only 5·1 per cent of the fishermen earned £150 or over per year and that 37·5 per cent earned less than £15 per year. They did not, therefore, have the capital to buy the equipment needed to catch the fish or process it in large quantities, and the stimulus of the Second World War was necessary before the capital became available. This is a parallel situation to that which prevailed in California, the First World War providing the stimulus in that case. The South African pilchard industry, therefore, dates from 1943. In South West Africa the industry is even younger, commercial fishing dating from 1947.

The pilchard is found along most of the 1500 miles of the west coasts of South and South West Africa and also along the south and east coasts of South Africa at certain times of the year. The hydrological characteristics of the area are such that at certain times of the year and at certain points along the west coast conditions arise which are ideal for the fish to form dense shoals. Areas in which dense shoaling is known to occur are the Baia dos Tigres in Angola, Walvis Bay in South West Africa and St. Helena Bay in South Africa. These form the basis of the fishery and it will be profitable at this point to study briefly the hydrological environment.

First, the conditions off the coast of South West Africa will be described. The northern and southern boundaries of this country are two large rivers, the Cunene River and the Orange River respectively. The distance between these two is about 800 miles, Walvis Bay, which is also the centre of the fishery, is approximately half-way between the two.

Winds near the coast tend to be rather variable but further offshore they are governed for most of the year by a high-pressure area which exists in the South Atlantic. South West Africa being on the east of this anti-cyclonic area has mainly southerly or south-easterly winds. This pattern is modified as the shore is approached, giving the more variable winds mentioned above. The winds have a bearing on the most important feature of the hydrology of the area, namely the upwelling of nutrient-rich waters from considerable depths and of Antarctic origin.

Peak upwelling in the Walvis Bay region is in late winter and early spring and is effected by the northerly flowing Benguela current which is the main body of water of Antarctic origin. As it travels north this current is deflected along the west coasts of South Africa and South West Africa. It continues in its northerly direction, but it also has a westerly moving component which is emphasized by offshore winds. As the surface waters move away to the west due to the current's own movement and the effect of the winds, the deeper waters are drawn from below to replace them. Poorly saline waters thus become established from the shore to beyond the continental shelf during winter and early spring, the main time of upwelling.

As summer approaches upwelling becomes less, so that at the end of summer and during the autumn upwelling has virtually ceased. This also correlates with the wind system because coastal winds are least strong during the summer. It is also found that there is considerable vertical stability during the summer as shown by vertical temperature and salinity measurements, and this increases as the summer progresses, reaching its maximum in February (the seasons being oppositely placed, relative to the months, in the southern and northern hemispheres). The stable conditions prevail until late autumn or early winter, that is until April, May or June. Normally, however, by early winter the highly saline water, found inshore during periods of stability, has begun to move offshore and is being replaced by the poorly saline water typical of the period of upwelling.

The rate of the upwelling water varies considerably from place to place and also with the stage of the process. For example, during the late winter, when the process is just beginning, water is probably rising at no more than a fraction of a metre each day. The rate of upwelling is probably at a maximum at about 1·9 metres in a day. Similarly, the depth from which the water upwells is variable, being from about 100 m down to slightly more than 300 m. It is interesting to compare this with the upwelling of California which is said to be from a maximum depth of less than 200 m (Sverdrup *et al.*, 1942). The area where water upwells from the greatest depths is off the Orange River. There, waters are reported to rise from more than 350 m depth whereas further north they rise from between 150 and 250 m (Stander, 1964).

As would be expected the seasonal changes in the process of upwelling also affect the salinity of the area. The changes in salinity, however, do occur with remarkable regularity and are broadly as follows. The water which is brought to the surface is partly at least of Antarctic origin and is, therefore, cooler and lower in salinity than other waters of the area. It is reported from the coastal observation stations of the South West African Administration's Marine Research Laboratory that the upwelled waters have a salinity of from 34·8 to 35·1‰. This order of salinity prevails until late summer and autumn when more highly saline water moves into the coastal regions with salinities of from 35·2 to 35·4‰.

In South African waters the conditions off the west coast are very similar to those just described for South West Africa, being the result of the same primary meteorological and hydrological conditions. The waters off the south and east coasts of South Africa are, however, of sub-tropical origin, being relatively warmer and more saline than those of the west. They are ascribed to the effects of a southerly travelling current, the Agulhas current, and do not provide the combination of conditions which give rise to large fisheries. More information is available concerning the hydrology of these two countries' coastal waters in Stander (1964) for South West Africa and Clowes (1950) for South Africa.

As already mentioned in considerable detail in Chapter 3, the presence of quantities of nutrient salts, particularly nitrates and phosphates, can be of considerable significance to the development of a fishery. These salts enable the development of phytoplankton which may in turn lead to a large standing crop of zooplankton. Both of these elements are important food for fish, especially for pelagic fish like the pilchard.

To give some indication at the outset of how these waters have been able to support a fishery which has expanded at what might be described as an alarming rate, Table 38 shows the landings of pilchards for the two countries separately. A total column is also given but

TABLE 38. QUANTITIES OF PILCHARD LANDED IN SOUTH AND SOUTH WEST AFRICA FROM 1947 AND 1951 RESPECTIVELY, THROUGH 1966
('000 metric tons)

Year	South Africa	South West Africa	Total
1947	30·0	—	30·0
1948	55·0	—	55·0
1949	70·0	—	70·0
1950	108·9	—	108·9
1951	99·4	127·2	226·6
1952	170·0	225·8	395·8
1953	137·0	262·3	399·3
1954	88·3	250·7	339·0
1955	122·0	227·2	349·2
1956	76·3	231·4	307·7
1957	107·5	222·8	330·3
1958	194·6	223·7	418·3
1959	260·2	271·3	531·5
1960	317·9	283·0	600·9
1961	402·4	342·3	744·7
1962	410·7	397·3	808·0
1963	400·9	546·7	947·6
1964	256·1	655·9	912·0
1965	204·1	666·0	870·1
1966	114·1	635·7	749·8

SOURCE: United Nations. Food and Agricultural Organization, *Yearbook of Fishery Statistics*, vol. IV(i), V, XI, XV, 20.

this must not necessarily be taken to mean that the same resource, biologically or geographically speaking, has provided all the fish. All landings available are given, and again these figures are in metric tons although the short ton is used in these two countries. Utilization of the catch has been into fish meal, oil and canned pilchards. The main markets for these products have had to be found abroad as the *per capita* consumption of fish in the two countries is very low, being about 10 lb of fish flesh per year compared with 17 lb per year in Great Britain (Culley, 1965), 55 lb per year in Norway or 70 lb per year in Japan (Irvin and Johnson, 1963). The fishing industries in these two countries have thus been built up on export trade, and this applies particularly to the pilchard industries where practically the whole of the pilchard fish meal and oil are sold to overseas buyers before it is produced and where probably 90 per cent of the production of canned pilchard is also sold abroad before the fishing season has ended. For example, in 1965 it was reported that the South West African factories had been concentrating on the production of fish meal and that the entire production of pilchard meal and pilchard oil had been sold in advance (*South African Shipping News and Fishing Industry Review*, 1965c). Of the three products from the pilchard-processing factories, canned pilchards, oil and meal, the emphasis on production of any one of these has been very different in the two countries. Although for most of the years since 1960 meal production has been the major product in both countries, in South West Africa it has frequently been nearly equalled by the output of canned pilchards. In South Africa, on the other hand, canned pilchards have invariably been third in importance, sometimes barely figuring in the statistics; as in 1959 when the production of canned pilchards was 0·4 per cent of the tota product output. However, the subject of catch utilization will be dealt with in more detail later.

The Biological Resource

The biological resource of the two industries is the fish known biologically as *Sardinops ocellata.* It will be recalled that the biological name for the California sardine is *Sardinops caerulea,* indicating close similarities for the two resources. Some authorities (e.g. Svetovidov, 1952), as mentioned in Chapter 2, have suggested that *S. ocellata* and *S. caerulea* should be classified also with three other species from Japan, Australasia and Chile as sub-species of *S. sagax.* A comparison of vertebral counts in the various areas in which *S. ocellata* has been caught has not suggested any racial differences within the species.

Food and Feeding

Many points of similarity in the diet and feeding habits exist between the South African pilchard and the other two examples dealt with. For example, Davies (1957) pointed out that it is primarily a filter feeder, straining large quantities of water through its gill rakers. Food caught on these is then transferred to the alimentary canal as in the other two species. It can, however, also be specific about certain items of its food and can exist for months in captivity by particulate feeding only.

Pilchards feed throughout the year but breeding members of the population may stop for a short period during spawning. Examination of stomach contents showed that most feeding occurred during the summer. Davies (1957) examined a large number (16,664) of pilchard stomachs before reaching his conclusions regarding the feeding and food of the pilchard. He showed that phytoplankton was the main food of the South African pilchard, being approximately twice as frequently eaten by volume as zooplankton. The actual mean ratio of the two types was 66:34, phytoplankton to zooplankton. However, this was not constant throughout the year as large quantities of phytoplankton were eaten in the autumn and spring when little zooplankton was eaten, but during summer and winter more zooplankton was taken. From his analysis of stomach contents Davies found that diatoms were the most important item of phytoplankton, and in fact the most important item of diet altogether. Table 39 shows the percentage of diatoms in the stomach contents for the four seasons.

Thalassiosira spp. and *Rhizosolenia* spp. are probably the most important of the diatoms and the latter at least also occurs as part of the diets of the other two species. As a brief comparison, the California sardine had diatoms as its most important item of diet in the summer, but the pilchard from the English Channel was shown to have diatoms as its most important item of diet during the spring and autumn.

TABLE 39. DIATOMS AS PER CENT STOMACH CONTENTS DURING SEASON FOR SOUTH AFRICAN PILCHARD

Season	Per cent
Spring	84
Summer	53
Autumn	67
Winter	55

Of the zooplankton recorded the copepods were the most important, and again this is a similar situation to that found with the other two species. Like the diatoms, representative species of the Copepoda were found in the stomachs throughout the year, but they were most frequently found in summer and winter. As with the other species *Calanus* and *Centropages* were among the most frequently recorded examples in South West Africa and South Africa. Figure 20.1 shows the percentages of the various groups eaten indicating their relative importance. It should be noted, however, that Davies (1957) thought that because the copepods were found throughout the year and because they were also quickly liquifiable, much of the 11 per cent classified as unrecognizable zooplankton would probably be copepod

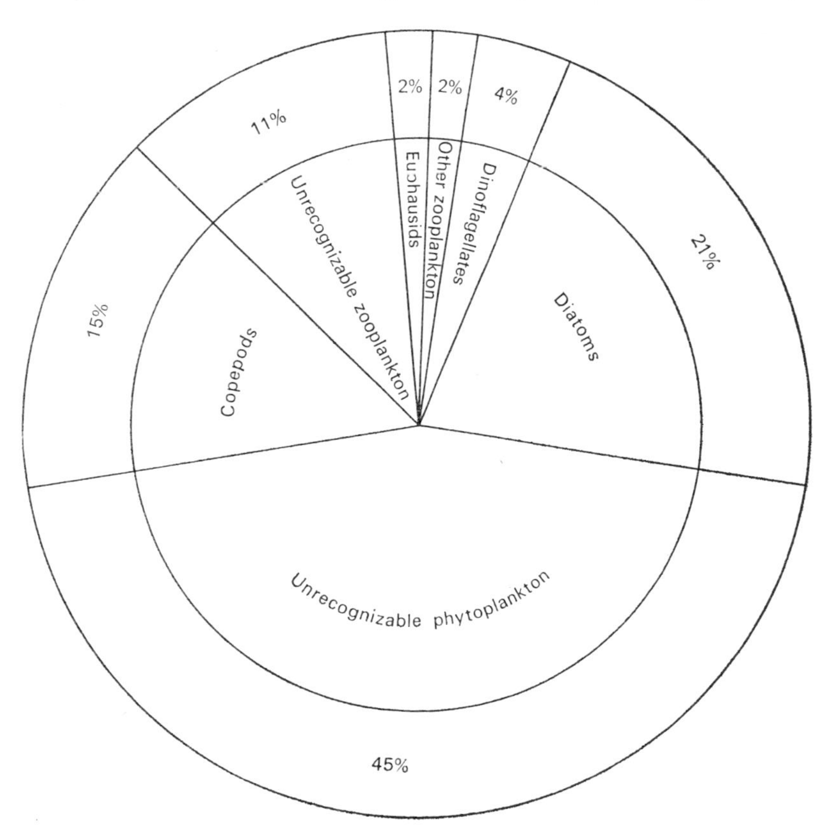

FIG. 20.1. Food of South African pilchard: percentage of various types eaten.

material. This would increase the percentage of copepod material to within the range 15–26 percent.

Breeding

Spawning occurs over a large percentage of the year as fish mature and move to the spawning grounds. However, as could be expected there are certain peak periods when spawning activity is intense. Off South West Africa there appear to be two main spawning seasons which are:

(i) the main period from August to October with a peak in September and October.
(ii) from January to March with a peak in February (Matthews, 1960a).

It has been found that it is usually larger fish which spawn in the first spawning period and the smaller fish which spawn in the second. Matthews (n.d.) has suggested that the pilchard off South West Africa do not spawn more than once during a year, but that they have a prolonged breeding season as a result of successive ripening of ova.

A similar pattern occurs off South Africa although de Jager (1960a) reported spawning to some extent throughout the year and Davies (1956a) stated that the more intense period of spawning may start in July and go on through the following April. This gives 2 months longer as a possible spawning period than is suggested for South West Africa. The peak activity within this period occurs during the same months as for South West Africa.

Sexual maturity is attained by some individuals when they are of 18·0–18·5 cm standard length, but it is not until they reach 20·5–22·0 cm standard length that all become sexually mature. Table 40 gives a more detailed length: sexual maturity analysis for *Sardinops ocellata* and this is also compared with some figures (based on Clark, 1934) for the female of *Sardinops caerulea.*

TABLE 40. STAGE OF SEXUAL MATURITY REACHED COMPARED WITH STANDARD LENGTH FOR MALE AND FEMALE SOUTH AFRICAN PILCHARD AND FEMALE CALIFORNIA SARDINE (Data from Davies, 1956a and Clark, 1934)

Sexual stage	Standard length (cm)		
	South African pilchard		California sardine
	Male	Female	Female
1. Immature	18·0 and under	18·5 and under	17·0 and under
2. 50 per cent mature	18·5–19·9	19·0–19·9	18·0–18·9
3. Small mature	20·0–21·9	20·0–21·9	19·0–21·9
4. Large mature	22·0 and over	22·0 and over	22·0 and over

An interesting biological phenomenon was observed by Davies (1956a) whilst he was making a study of the gonads of mature and immature fish. He found that the gonads of mature fish show a weight cycle which exhibits peaks around November and January with troughs between these periods. This sort of pattern could well be expected, knowing the main spawning seasons, but he also noted that even the gonad weights of the immature pilchards showed the same weight cycle, suggesting that there was an innate rhythm connected with the sexual physiology of the fish which was not dependent on the full hormonal complement being present.

As with other related species the range of water temperature in which eggs of *Sardinops ocellata* are found is over several centigrade degrees although the majority of eggs have been found within a range of only 4 degrees. For example, Davies (1956a) recorded that 78 per cent of the eggs tend to be found between the temperatures of 13·0° and 17·0°C, but that eggs have been found between 11° and 22°C. This range is considerably greater than for the California sardine, although the range of maximum spawning is similar in that species (a 3½-degree range from 12·5° to 16°C). The pilchard of the English Channel also is thought to breed mostly within a 3 degree range but in that species the optimum is lower, being from 10°C to about 13°C. It is worth mentioning that different papers concerning total and optimum spawning temperature ranges invariably differ, sometimes considerably even when discussing one species. For example, de Jager (1960a) broadly agrees with Davies (1956a) but gives the following figures. Overall range 11·9–21·7°C; optimum spawning range 13·8–16·0°C. Matthews (1960a), referring to waters off the coast of South West Africa, gives 11·2–21·2°C as the overall range in which pilchard eggs have been recorded and 14·9–19·6°C as the optimum range.

Spawning of the South African stock seems to take place both further offshore and to the south of the commercial fishing grounds (Davies, 1957a), sexually mature fish with ripe and running gonads being caught very infrequently on the commercial grounds. This is also the case for the South West African stock and there the fish are also thought to move further offshore than the commercial fishing grounds. Eggs are found in considerable quantities as far as 250 miles (400 km) north of Walvis Bay and in a band from 20 to 55 miles (32–87 km) offshore (Matthews, 1960a).

Davies (1956a) gives an estimate of the number of eggs spawned by the female of *Sardinops ocellata* as 95,000. He found that a fish of standard length 23·3 cm had just over this number of ova, although larger fish had rather more ova. If a female only spawns once between July and April then the maximum possible is in the region 90,000–100,000. However, if the two peaks in the gonad weight cycle indicate two spawnings, then between 180,000 and 200,000 eggs spawned per season would be possible. It will be recalled that this was mentioned as a possible number for the California sardine also, although the British pilchard probably lays no more than about 70,000 eggs per year.

Development

The egg possesses the characteristics outlined for the other two species and the rest of the development also follows broadly similar lines. It should be mentioned that the young pilchard of about 40 mm length, sometimes called the juvenile, does not possess scales but otherwise resembles the adult.

Predators

The egg of the pilchard, being planktonic, suffers many losses from a variety of predators. These range from pelagic molluscs (de Jager, 1960a) to microphagous fish. Generally, any organism which eats plankton will probably take pilchard eggs at some stage, as will adult pilchards themselves.

Larval and post-larval stages are preyed on by the squid *Loligo reynaudii*, fish such as snoek (*Thyrsites atun*) and maasbanker (*Trachurus trachurus*) and sea birds. These latter are also the most important predator, apart from man, on the adult stages. On the islands off the coast of the Union of South Africa there are three important predators as sea birds and these are: (1) the Cape cormorant (*Phalacrocorax capensis*), (2) the Cape gannet (*Sula capensis*) and (3) the Cape penguin (*Spheniscus demersus*). Davies (1958a) estimated the annual pilchard consumption of each of these species. His estimate was based on several variables making the actual figure arrived at somewhat tentative. It, nevertheless, is interesting to obtain an idea of the order of magnitude of this natural predation.

Some of the variables mentioned above are given below.

(a) An estimate of the total population of the three types. This was carried out by air and is presumably open to a considerable margin of error.

(b) The number of feeding days per year was arrived at rather vaguely with numbers of days subtracted from the maximum possible of 365. Causes which might contribute to the non-feeding of the birds were: (i) times when the birds may have migrated from the area, or (ii) when meteorological conditions such as fog are likely to have made fishing impossible, or (iii) days during which fish would not be available.

It was suggested that the birds would feed, therefore, for only two-thirds of the year, that is 244 days.

Davies's final figures are given in Table 41, and included are those he gave for consumption of the maasbanker (or horse-mackerel, *Trachurus trachurus*) as the two fish are often caught and processed together in South Africa.

TABLE 41. ANNUAL CONSUMPTION OF COMMERCIALLY IMPORTANT FISH BY THE SEA BIRDS OFF THE COAST OF THE UNION OF SOUTH AFRICA (SHORT TONS)
(From Davies, 1958a)

Sea bird	Pilchard		Maasbanker	
	All sizes	Commercial sizes	All sizes	Commercial sizes
Cape cormorant	6,000	4,000	1,000	250
Cape gannet	16,000	12,500	4,800	3,600
Cape penguin	21,300	9,200	980	—
Total	43,300	25,700	6,780	3,850

Davies points out that the difference between the consumption in commercially sized fish and that of all sizes consists largely of late juvenile pilchards. This is a period when natural mortality is comparatively low so that the actual effect on the fishing is more likely to be represented by this larger figure; that is just over 50,000 short tons per year of pilchard and maasbanker combined (see also Fig. 20.2).

A similar estimate was made for the sea bird population off the coast of South West Africa but figures of overall consumption for the three genera could not be arrived at as no aerial survey giving total population estimates had been done. Matthews (1961) considered the three types covered by Davies and concluded that the Cape cormorant and gannet between them consumed very considerable numbers of pilchard. A tentative range of consumption for the cormorant alone based on varying estimates of the population in the Walvis Bay area was 32,625–65,000 short tons per year. The Cape penguin, however, was found not to be an important predator, in South West African waters, very few being present in the Walvis Bay area.

Other predators on the adult pilchard are, amongst fish, the snoek, which is known to prey heavily on pilchard shoals when they are available; the stockfish, which attacks them

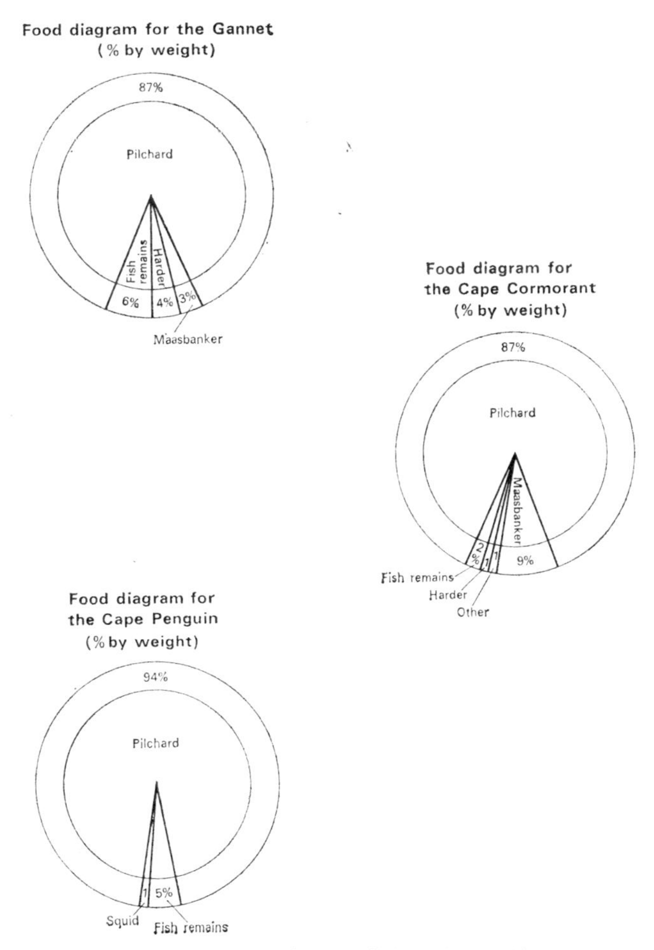

FIG. 20.2. Consumption of pilchard by sea birds.

on their spawning grounds; and, along the east coast the Spanish mackerel. The names for these three species are *Thyrsites atun*, *Merluccius capensis* and *Scomberomorus commerson* respectively. Davies (1956b) also noted that the geelbek (*Atractoscion aequidens*) preys on the adult pilchard off the south coast of South Africa. Amongst mammals, other than man, predators are known to include the cape Fur Seal, *Arctocephalus pusillus*, and porpoises.

Parasites

Both South African and South West African marine biologists generally report that *Sardinops ocellata* is a very healthy fish, being remarkably free from parasites and diseases.

It does, however, suffer from a parasite, the protozoan *Chloromyxum thyrsites*, which earns its specific name from the snoek in which it is also prevalent. This does not seem to have any effect on a living fish but it has a serious effect after death as the infected fish are usually unsuitable for canning and, as mentioned in Chapter 12, become "mushy". According to Dreosti *et al.* (1957) the parasite secretes an enzyme which in a living fish is removed from the tissues, almost immediately after its production, by the circulation. After death, however, the enzyme will remain and not be removed or broken down by defence mechanisms of the host. Eventually it will diffuse through the tissues of the host and break down all the muscular tissues.

If an infected fish is canned then its flesh will turn mushy and will break up when being taken out of the can. Unfortunately, the presence of the parasite cannot be detected until several hours after death or until it has been cut open when the following two methods can be applied to ascertain whether infection has occurred.

(i) If a cut fish which has been dead 5 or 6 hours has pressure applied to it, pinkish matter will appear between the muscles at the cut ends if it is heavily parasitized.
(ii) An infected fish usually has "white spots or pustules" beneath the lining of the somatic cavity which are caused by dense aggregations of *Chloromyxum* spores. The number of these give "a rough guide as to the extent of the infection" (Dreosti *et al.*, 1957).

With infected fish there are two factors, both of which have very broadly defined limits, which determine whether or not they will be suitable for canning. These factors are: (a) the actual number of spores in the flesh and (b) the time between catching and canning. It was reported by Dreosti *et al.* (1957) that in 1955 when infection with *Chloromyxum* was particularly heavy 6–15 per cent were definitely unsuitable for canning but another 25–30 per cent were borderline cases. It would seem to be important to find some more definite and accurate method of determining whether or not fish will be suitable for canning as even borderline cases are not suitable for human consumption even though the parasite is apparently harmless to man.

Migratory Movements

The broad migratory movements of the South African pilchard do not seem to be fully known. This is unfortunate as these movements could well affect the interrelationships of the fisheries in the two areas, South Africa and South West Africa. Until about 1960 it was assumed that the same resource was the basis for both fisheries. Davies (1956a) stated that the size composition and reproductive activity of the two groups suggested that they are interrelated. This being so, he pointed out, extensive overfishing by the South West African boats could lead to the downfall of the stock of the Union.

In an attempt to determine the migratory patterns of the pilchard and the extent of the intermingling of the fish from South African and South West African waters, a joint, extensive tagging programme was started in November 1956. Although full results do not appear to be available, those that have been published indicate that in spite of some mingling of fish from the two areas, on the whole this is of only a very casual nature. It would appear that the fisheries can operate independently without overt fear of too much fishing in one area affecting conservation measures in another. One hundred and seven thousand fish had been tagged in the Walvis Bay area and 38,000 off the Cape West Coast. Between 1957 and 1960 only three tags had been recovered. These were recovered from fish caught in the St. Helena Bay area and tagged in Walvis Bay. Then in 1961 twelve tags from Walvis Bay fish were recovered in Union factories and, as the leading fishing journal of South Africa reported, "[this] has proved reasonably conclusively that there is some mingling of South West African and Cape pilchards" (*South African Shipping News and Fishing Industry Review*, 1961b). It would seem, however, that although this "mingling" undoubtedly takes place it is only a very small part of the population normally resident off South West Africa which makes this southward migration. Furthermore, it does not appear to be an annual migration, no tags at all being recovered in 1959 and also 1960, even though improvements had been made to the recovery system.

No tags from fish tagged in Union waters have been recovered in Walvis Bay factories, suggesting that there is no long northward migration. The number of tagged fish released in the south was many fewer, but if there was any consistent northward movement it is likely that at least one or two South African tags would have been recovered in South West Africa.

At the moment, therefore, the results of the tagging programme tentatively suggest that there are occasional southerly migrations of larger pilchards from South West African waters. No movement on a large scale in the opposite direction, however, would appear to take place. Stander and le Roux (1968) state categorically that the experiments show "a definite annual influx of northern fish from South West Africa, but there is no evidence of larger-scale movements of fish between the two areas".

Regular movements of fish within the area of waters off the two countries do occur and are comparatively well documented. Matthews (1960a) stated that spawning fish, which are rarely found in the commercial catches, move from the grounds which are commercially exploited to the areas outlined previously as spawning ground, i.e. to the west and north of Walvis Bay. This is also the case off South Africa where the breeding takes place in waters further to the west and south of the commercial grounds. The onshore–offshore movements off South Africa can be summarized as follows.

In the spring (September, October and November) catches close inshore are high. Also catches made from between 5 and 10 nautical miles from the shore are very high at this time, but they begin to decline during October and during November decline to nothing.

During the summer months (December, January and February) inshore catches begin to increase again from nothing in December to considerable numbers being caught in late January and early February. Later in February offshore catches are considerable. That is, more of the fish are now found more than 10 nautical miles from the shore.

Autumn brings the beginning of movement from offshore waters to those inshore. In March the greatest catches are taken more than ten nautical miles from the coast, but in April catches in the 5–10 mile zone increase indicating that the movement toward inshore waters has begun. In May fish again appear inshore and catches between 5 and 10 nautical miles offshore reach their maximum.

In early and mid-winter, June and July, the greatest number of catches are made close inshore with offshore catches at a minimum. During August, however, catches in the 5–10 mile zone and beyond increase.

Occurrence of these regular movements in both South and South West African waters is a further indication that the pilchard population of these two areas are largely self contained. This is further supported by the fact that factories in the Walvis Bay area recovered many tags from fish tagged in the area. Furthermore, the ova and the larvae tend to be transported inshore from the spawning grounds by currents and are retained very much within the general area of the country nearest to which they were spawned. For example, for South Africa, it is known that the nursery grounds for immature pilchards are areas inshore between Lambert's Bay and False Bay. It appears unlikely that there can be any mass migration from either fishing area to the other.

A final problem connected with the migratory movements of the pilchard concerns the presence of pilchards along the south and east coasts of South Africa from Cape Point as far as Durban. No commercial fishing for pilchards occurs in this area but small, mostly immature fish are found inshore along the east coast for about three weeks in the year and up to about 100 metric tons may be caught by beach seine (du Plessis, 1958). These fish migrate north-east along the coast as far as Durban and then disappear. It is thought that they return to the Cape in deeper waters further offshore, aided by the Agulhas current. No explanation can really be offered either for this or for the, at least occasional, occurrence of juvenile pilchards in the same area as suggested by the presence of juvenile pilchards in the stomach of a shad (*Pomatomus saltater*) caught at Durban in November 1955 (Davies, 1956b). There is no apparent connection between these movements and the two usual migratory stimuli, feeding and breeding (Davies, 1957a).

CHAPTER 21

THE POPULATION IN RELATION TO THE FISHERY

Structure of the Population

Sex Ratio

It is only from the commercial catches that the sex ratio of the population of a fishery resource becomes known. In the shoals not sampled by the commercial fishery there may be slightly different sex ratios. However, in both South Africa and South West Africa the ratios are remarkably similar, being in South Africa, male:female = 44:56 (de Jager, 1960a) and in South West Africa, male:female = 40:60 (Matthews, 1960a). These are, of course, corrected figures and the percentage of the sexes in the catches may vary up to about 5 per cent per year as shown in a paper by Matthews (n.d.). This gives the percentage of males in the commercial catches over the 4-year period from 1957 through 1960 and Matthews's figures are reproduced in Table 42.

TABLE 42. PER CENT MALES IN COMMERCIAL PILCHARD CATCHES OF SOUTH WEST AFRICA

1957	38
1958	40
1959	40
1960	35

As well as this variation in the ratio for the total commercial population there are variations over different size ranges. These differ considerably in detail between the two countries, but the overall tendency is the same. That is, in the lower length groups males tend to be more common than females, but with increasing size males become less frequent so that by the time 22·0–23·0 cm standard length has been reached the commercial stock is almost entirely females.

Age and Size Composition

Figures for the age and size composition of the commercial catches for the two countries are not available for 1959 onwards as yet. In relation to the age of the fishery, therefore, a large percentage of the years is undocumented and it is probably since that time that most significant changes in the age and size structure of the population will have occurred.

Even by the end of the 1958 season Nawratil (n.d.) pointed out that the number of fish caught aged IV+ had decreased steadily from 1952 through 1958. There had been a similar trend, though not so consistent, for the III+ age group and an opposite trend in the I+ age group. This is clearly illustrated in Table 43.

TABLE 43. PER CENT'S OF DIFFERENT AGE GROUPS IN SOUTH WEST AFRICAN COMMERCIAL CATCHES, 1952 THROUGH 1958

(Figures from Nawratil (n.d.))

Year	Age group			
	I+	II+	III+	IV+
1952	2·7	41·4	45·5	10·4
1953	3·9	45·0	43·7	7·4
1954	2·2	46·3	46·7	4·8
1955	1·5	44·1	51·1	3·3
1956	6·5	43·1	47·1	3·3
1957	9·5	47·4	40·3	2·8
1958	13·6	55·5	29·2	1·7

It can be seen from these figures that between 1956 and 1958 the percentage of I+ year old fish in the commercial catches increased markedly in what appears to be the beginning of a new statistical trend. As the total annual catches of pilchard were at this time limited to 226,900 metric tons per year, although in practice they were always somewhat above this, it means that these figures indicate a real increase in the number of younger fish caught. This is not necessarily bad fishing practice as pointed out before, but if continued to extremes would mean that best use of the resource was not being made. This is because with young fish the growth curve is at its steepest and the transformation of food assimilated into flesh is at its greatest (Fig. 21.1). This is the part of the food chain in which man is interested and it is bad economic practice to cut it off too severely half-way through its most profitable stage. It is possible to look at the fishing of the lower age groups in another way also. Matthews (1960b) stated, although as became evident in the discussion of the California population studies not every one would agree with him:

> In the case of small fish the intensity of fishing operations naturally has a considerable influence on the quantity of large fish which will be available in the fishery at a later stage. The fewer large fish there are in the total population, the smaller the number of fish which can reproduce. This will result in poorer breeding years which will in turn lead to a smaller increase in the number of young fish, and ultimately the whole fishery will be adversely affected.

The above quotation appears to be stating definite facts, but it is really posing a problem, the controversial nature of which has occupied biologists for years. For a fuller treatment of the ideas the reader is referred to Chapter 18.

Table 43, giving the age composition of the South West African commercial catches, indicates that between 85 and 95 per cent of the catch was drawn from the II+ and III+ age group during the years 1952 through 1958. Nawratil (n.d.) stated that II+ and III+ year old fish had average total lengths of 19·5 and 25·0 cm respectively. Converting the percentage of the II+ and III+age groups to length groups, therefore, for the 7 years under consideration there was an annual average of 46 per cent of fish having an average length of 19·5 cm and 43 per cent having an average length of 25·0 cm. This left 11 per cent each year to be divided in varying proportions to the I+ and IV+ age groups which would have average total lengths of 12·9 and 27·9 cm respectively. It can be seen from the figures in the table how far 1957 and 1958 had begun to deviate from the previous pattern. It is almost certain that any new figures would give a completely different age and length composition of the commercial catches of South West Africa.

The South African fishery, from figures given by du Plessis (1960), affects the commercial stocks in a rather similar manner to the South West African fishery. No one or two length groups are as heavily represented, but between 1952 and 1957 inclusive the majority of the

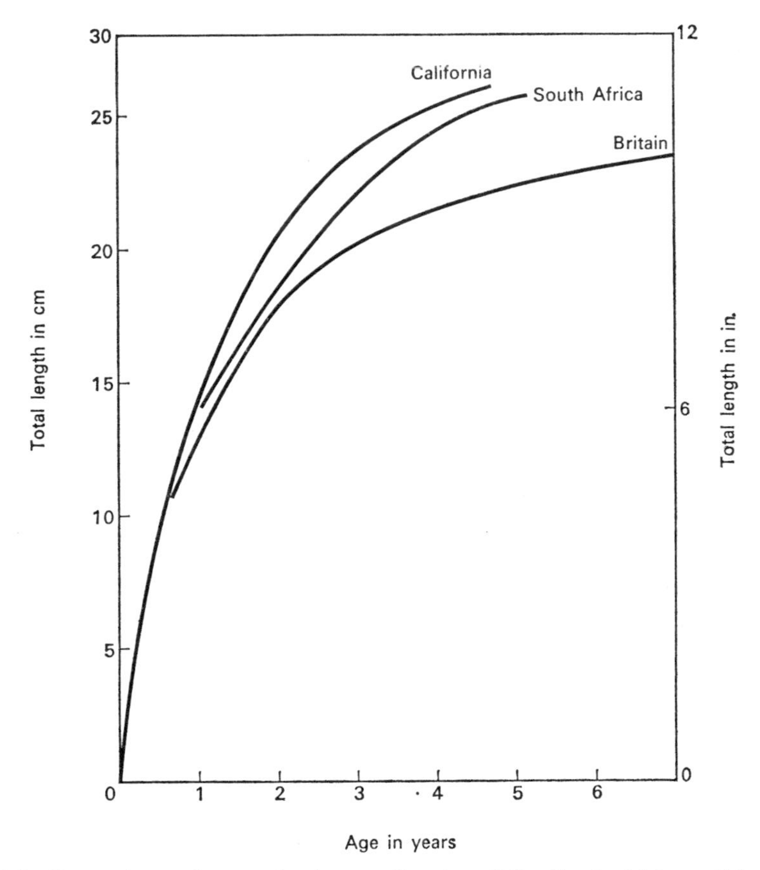

FIG. 21.1. Comparison of approximate growth rates of the South African pilchard, the California sardine and the British pilchard.

catch consisted of fish between the standard lengths of 21·5 cm and 23·5 cm. Total length measurements are not available, but from this it can be seen that the fish caught off South Africa were larger during this period than those caught off South West Africa. In 1958 the catch consisted largely of fish in the 18·0–20·0 cm standard length groups. This was possibly exceptional as du Plessis (1960) stated that there was an unusual number of smaller pilchards in the commercial fishing area during that year.

The larger size of the pilchards caught off South Africa, when compared with those caught off South West Africa, would appear to be due to the fact that the South African fish are older. The figures quoted from Nawratil (n.d.) for South West Africa clearly indicate that the dominant age groups between 1952 and 1958 were II+ and III+ year olds. Comparable figures are available for South Africa between the years 1952 and 1955 (Davies, 1958b). These show that the dominant age groups were from III+ to VII and are given in Table 44.

TABLE 44. DOMINANT AGE GROUPS FOR MALE AND FEMALE PILCHARDS IN THE SOUTH AFRICAN COMMERCIAL CATCHES FROM 1952 THROUGH 1955

(Figures from Davies, 1958b)

Year	Male	Female
1952	III to VI	III to VII
1953	IV to VII	IV to VII
1954	IV to VI	IV to V+
1955	IV to VI	V to VII

From these figures there is no indication that up to 1956 the South African fishery had been fishing out the older age groups, and this is confirmed by a recent paper on the problem (Stander, 1965). The impression given by this paper is that pilchards in age groups similar to those given in Table 44 were caught until 1957. It is suggested, as a possible explanation of this phenomenon, that smaller fish were rarer in the stock before 1957, and particularly during 1954 and 1955, due to poor recruitment. During 1957 a new year-class appeared in the fishery. This was the II+ year-class of 16 cm length. Since then at various times the younger, smaller fish of 15–17 cm length have played an important part in the fishery as indicated by the following figures. The 1962 catch of pilchards in South Africa was a record for the industry which still has not been surpassed. This catch was nearly four times that taken in 1950 but, as Stander (1965) noted, six times as many fish were required to make up the 410,700 metric tons.

Since 1962 pilchard catches for South Africa have gone steadily downwards so that the 1966 total was only 114,100 metric tons. This suggests that the pilchard stock off South Africa has now had the II+ and older age groups so heavily fished that there is at present no reserve of fish to compensate for years of poor recruitment to the fishery. This is what happened to the California sardine fishery in the mid-1940's. There, however, the decline was a rather fluctuating process, with a large fleet applying massive fishing effort to drag as much of the available stock as possible from the sea thereby doing it very great and long-lasting damage. The latest developments in South Africa such as the refitting of a former whale factory ship to catch pilchards outside the 12-mile limit may indicate the beginnings of similar activity in that country. Such a ship would be free from quota and licensing restrictions and in this way is a forceful reminder of the damaging law-evasion of the California floating plants during the 1930's. More will be said concerning this project later.

Recruitment

Recent figures for the abundance of eggs in the waters off South Africa are not available. Du Plessis (1960) gave figures for the period 1952 through 1958. It was thought that there might be some correlation between the abundance of pilchard eggs in one year and the availability of the fish, as expressed by the size of the commercial catch, 4 years later. There was some correlation along these lines for the years 1952, 1953 and 1954 with 1956, 1957 and 1958. Since then, however, no correlation exists as the catch rose from just under 200,000 tons in 1958 to over 410,000 tons in 1962 whereas recruitment, as expressed by the abundance of eggs, remained fairly steady and low. Obviously, the massive increase in effort required to obtain such a rapid increase in the size of the catch would not give the stock time to adjust itself to the new conditions. Consequently, no relationship between recruitment and availability could be shown to exist. Further, if the relationship already noted between the abundance of eggs during 1952, 1953 and 1954 and the catch in 1956, 1957 and 1958 was an indication of a working relationship for the South African stock, then the catch should have stayed between 100,000 and 200,000 tons per year at the highest. The steady decline since 1962 may be an indication that damage was done to the stock by allowing the catch to soar to over 400,000 tons per year.

Further information on recruitment is available in recent articles by Stander (1965) and Stander and le Roux (1968). It seems that the level of recruitment for 1958 continued through 1960. The following 2 years suggested a considerable increase, but in 1963, 1964 and 1965 it dropped to a level lower than 1960 as indicated by sampling the juvenile population by blanket net. Ratios comparing the size of pilchard recruitment to that for anchovy indicate that the anchovy is gaining ground at the expense of the pilchard, at least for the period 1958 through 1965. Table 45 gives some idea of the increasing ascendancy of the anchovy. This is a very important trend the significance of which will be mentioned again later.

TABLE 45. RECRUITMENT OF PILCHARD COMPARED WITH ANCHOVY AS INDICATED BY THE RATIO OF THE TWO FISH IN BLANKET-NET SAMPLING, 1956 THROUGH 1965

(From Stander and le Roux, 1968)

Year	Ratio	
	Pilchard	Anchovy
1956	15·0	1·0
1957	12·0	1·0
1958	2·0	1·0
1959	2·0	1·0
1960	41·0	4·0
1961	1·0	1·6
1962	1·5	1·0
1963	1·0	9·0
1964	1·0	19·0
1965	1·0	8·0

Conservation and the Pattern of Landings

It has already been pointed out that these largely post-Second World War industries have expanded at a very rapid rate. The South African landings showed a steady increasing trend from 1947 to 1952 and again from 1956 to 1962. It is possible that the fluctuating catches between 1952 and 1956 were at a rather low level because of the conservation measures imposed on the industry by the government. Of the various control measures used, however, it is considered (Stander, 1965) that the only one to have a limiting effect on the catch would be the gross hold capacity of the fishing fleet which stood at around 8320 short tons (7550 metric tons) between 1953 and 1960. Nevertheless, it seems likely that even this measure did not have much effect as the 8320 short ton hold capacity was sufficient to enable over 250,000 metric tons to be landed in 1959 whereas the average size of the landings from 1953 through 1956 was only 103,000 metric tons. The fluctuations in landings during the early period of the development of the industry probably reflect availability of the resource. As in all pelagic fisheries, fluctuations in availability would occur and could not, in the early days of the fishery, be overcome. Later, as the experience of the fishermen widened and there was more equipment introduced into the catching side of the industry, the fleet became generally more efficient and increased the proportion of the stock available to the fishermen. This probably accounts for the steady increase in landings which occurred in South Africa from 1956 through 1961 and which continued in a less marked manner through 1962 also. The fact that the catching side of the fleet became so efficient, landing over 400,000 metric tons in South Africa during 1961, 1962 and 1963, makes the marked drop in landings in that country between 1964 and 1967 all the more ominous.

The South African Administration started, after the first 3 years of fishing for pilchards to impose certain measures on the industry to control the catches. The example of the decline of the pilchard industry in California was salutory to them and there was a determination that if at all possible, the same thing should not happen in South Africa. The restrictive measures imposed and the years in which they came into operation are given below:

(i) The provision of a minimum mesh size for pilchard fishing nets (1950).
(ii) The fish meal plants were limited in number and in their reduction capacity (1950).
(iii) Fishing for pilchards was prohibited during certain months of the year (1951).
(iv) The number of boats allowed to fish for pilchards came under strict surveillance (1953).
(v) The number of licences issued for canning pilchards was limited (1953).
(vi) Catch quotas (for pilchards and maasbanker combined) were first introduced in 1953.

The purpose of each of these measures is fairly evident, but briefly they and the effect they might have are described below:

(i) The minimum mesh size was to prevent sexually immature fish from being caught so that the future breeding power of the population would not be reduced. All fish should theoretically be able to breed at least once in life. As was seen in California, however, this does not necessarily allow a population to maintain itself against commercial and changing environmental conditions.
(ii) The purpose of limiting the reduction potential of the factories was to prevent too many fish being taken from the sea by uncontrolled demand from an unlimited number of different commercial concerns. Although the existing reduction capacity is not worked to the full and it might appear that this measure has had no effect, it is very likely that it has been useful in holding back the development of too great a fishing effort which would have arisen if a greater number of commercial interests had been involved.

(iii) The closed season was introduced to coincide with the breeding season and has possibly been very useful in allowing breeding to take place comparatively unmolested, although in practice the fish disappear from the commercial fishing grounds during the breeding season.

(iv) Limiting the number of boats allowed to fish for pilchards has had the effect of limiting the hold capacity of the fishing fleet and as a consequence of this there may have been a limiting of the quantity of fish caught. However, the efficacy of even this measure is not certain, as pointed out above. The reason for this is that, with the increase in vessel size which has occurred over the past years, there has been an increase in hold size per vessel and no official control over the aggregate hold size of the fishing fleet. In 1953 as Stander (1965) pointed out there were 229 vessels in the pilchard fleet having a total hold capacity of 7550 tons. By 1965 the number of vessels in the fleet had dropped to 122 but the total hold capacity had risen to 10,800 tons; that is, there had been a fall of nearly 47 per cent in the number of fishing units, but their hold capacity had risen by 43 per cent. In another paper on this subject (Stander and le Roux, 1968) it is stated that between 1950 and 1965 the total hold capacity of the South African pilchard fleet rose by about 70 per cent.

(v) The restriction of canning licences did not exercise much influence on the catching side of the industry in South Africa as a canning plant on its own cannot survive in competition with a factory where canning and reduction occur together.

(vi) Probably the most fundamental conservation measure that can be imposed on a fishery is a catch quota. It is, however, an extremely difficult problem to decide at what figure this quota should be fixed. In South Africa a limit of 250,000 short tons was set in 1953 (that is equivalent to 226,860 metric tons). However, as soon as the landings approached this figure in 1959 and 1960 the quota was abandoned, the catches in those two years being 260,000 and nearly 320,000 metric tons respectively. It was unfortunate that at this time Peru's production of fish meal had flooded the world market and the value of the South African production had fallen considerably, prices in mid-1960 being from as much as 30–50 per cent lower than in mid-1959. (*South African Shipping News and Fishing Industry Review*, 1960.) This, added to the fact that the fish in 1960 were very rich in oil as well as being plentiful, made it appear that the resource would stand a very much greater exploitation than previous catches had hitherto led the authorities to believe.

The industry used all these factors to press for the temporary abandonment of the 250,000 short ton limit and since this was granted the limit has never been re-imposed. In 1964 and 1965 the catches fell to 256,000 and 204,000 metric tons respectively; a case of the resource imposing its own limit.

In South West Africa the 1951 catch, which was the first significant one, was about 125,000 tons and it rose rapidly so that by 1953 it had reached about 260,000 tons. In that country also a close watch has been kept on the fishery in an attempt to prevent overfishing occurring. After 3 years' fishing, therefore, a catch limit was imposed so that each of the four factories operating at that time could not process more than 42,544 tons in any one season. The two new factories starting operations that season were also given a limit, rather lower, at 28,363 tons. At the inception of the industry in that country it should be noted that the number of plants allowed to reduce pilchards was limited to six. The reduction capacity per hour for each of these factories was also controlled. A closed season of 3½ months between mid-November and the end of February was also in operation, but this did not seriously affect catches. The quotas were generally easily reached during the remaining part of the year during which time the oil content of the fish is higher. The closed season was abandoned in 1959–60. Mesh size was another factor controlled by the South West African Administration. Between 1949 and 1956 it was a 19-mm bar minimum. In 1957, however, with the general acceptance of synthetic fibres it was reduced to a 16-mm bar measurement. In practice, this reduction had little effect as cotton nets gradually shrink, so that in time the mesh would be reduced to about a 16-mm bar measurement.

The overall catch quota set in 1953 which totalled 226,900 tons was in operation for some years, and the graph showing the landings of South and South West Africa (see Fig. 22.1) indicates that the level of the catches was brought down to and held at this level over a period of 5 years. In 1960 the quota was raised slightly to 271,937 tons and in 1961 a new limit of 375,000 short tons (340,290 metric tons) came into operation. It was stated that this quota was to be divided equally among the six factories and that it would not be changed "unless scientific advisors to the Administration believe that the condition of the fishery and the fish resource justifies it". (*South African Shipping News and Fishing Industry Review*, 1961a.) The limit for 1960 was exceeded in that year, the catch being over 280,000 tons, so that it was a continuation of that trend which enabled the new limit in 1961 to be reached immediately.

In 1962 it became apparent that the limit was not being based on any scientific work of a long-term nature as the limit was raised again to reach 394,736 tons. Referring back to the previous quota increase it must be safe to assume that the "scientific advisors to the Administration" were quite happy about the long-term effects of such an increase. Furthermore, it must be assumed that they continued to be happy as the quota was raised twice in 1963 and again in 1964. Of the 1963 increases the second, of 10,000 short tons to each factory, was to be only temporary. It brought the allocation for each factory to 100,000 short tons (90,744 metric tons) and the total quantity permissible to 600,000 short tons (544,664 metric tons). The increase of the 1963 over the 1962 quota was nearly 150,000 tons. A further 108,693 tons was divided among the six factories in 1964 bringing the total quota to 635,357 tons (720,000 short tons). No further increase in the quotas occurred until mid-1967 when five of the eight pilchard factories were allocated another 9600 short tons (8711 metric tons) bringing the total to 768,000 short tons (696,914 metric tons). (*South African Shipping News and Fishing Industry Review*, 1967g.)

It can be seen that as a means of controlling the catches a quota system can only be effective if sufficient time be allowed between increases to enable the effect of the new quota on the stock to be assessed. However, without the quotas for each year it is possible that the factories would have processed even more fish and it must be acknowledged, therefore, that the imposition of the quota may have had some steadying effect on the development of the industry. On the other hand, the very existence of a quota introduces an element of competition to the fishing scene which may act as a continual spur to further activity until the goal is reached. In fairness it must be pointed out that in 1965 and 1966, in spite of requests for further quota increases, the Administration resisted pressure and the quota remained at the 1964 level, and it is at such times that the quota can be seen to be fulfilling its function. In practice, the quota was slightly exceeded in both 1964 and 1965, but presumably the catch was not as great as it might have been had there been no limit. The 1966 catch was at the level of the limit of the quota. One difficulty of a quota system of fishery regulation is that it must inevitably be rather arbitrary until a great deal is known concerning stock levels and recruitment. It was a good idea of the South West African Administration to fix a quota in 1953. It was also a good idea to raise it in 1960 when it became apparent that the stock was having no difficulty in maintaining the 226,900 ton level. However, it seems that the increases between 1960 and 1964 have been too rapid and astronomical, as neither the stock nor the scientists could possibly have been able to adjust to changing conditions so quickly. However, another difficulty about an arbitrary quota system is that it is possible to look at the graph for the landings in South West Africa (Fig. 22.1) and say that the stock can obviously stand the level of exploitation to which it is presently subjected, there are no signs from the statistics, or even the fishermen, that the stock is being overfished. Indeed, it was only the imposition of the 1953 quota which held back the development of the industry: thc 1965 level should have been reached in 1960. All this is surely a possible interpretation of the facts. It is hoped that it is correct, but it is possible that this rapid plunging into the unknown by the South West African Administration may have serious and far-reaching effects upon the pilchard industry. In fact, although the fishermen do not have great difficulty in finding enough fish for the factories to fulfil their quotas, there are now signs that the stock may be starting to suffer under the fishing effort imposed upon it. For example, the larger number of fish necessary to obtain a given weight of pilchards in the 1963 catch.

There has also been in operation a gross tonnage limit for the factories' boats, but this has not had any appreciable effect on the quantities of fish withdrawn from the sea, although it may have had some effect on the running of the factories.

It should be remembered that the sardine industry of California was able to withstand a very high level of exploitation for 11 or 12 years. After that, with increasing effort, the catch was maintained with fluctuations between 110,000 and 370,000 tons per year for 7 years. Since then the only predictable feature of the catches off California has been that they will be below 100,000 tons, and for the past 5 years they have been below 10,000 tons per year. (In Baja California greater quantities, rarely more than 20,000 tons per year, are still caught.) The California level of exploitation was very high, the South West African level is even higher and, if it is permissible to draw comparisons, it may be that the stock will not be able to withstand such exploitation for much longer: landings of South West African pilchard have been above 500,000 tons per year for 5 years already.

To summarize, it would appear from the South African landings and pilchard recruitment figures that the stock has been overfished so that a reappraisal of pelagic fishing activity there is needed. In South West Africa where landings remain at a very high level the stock is showing signs of being fished right down so that annual fluctuation in landings will probably soon start to occur. The increasing number of small fish in the catches (see Table 43) suggests that the industry there will shortly be relying on the latest year's recruitment to the available stock to sustain the fishery. This, as illustrated by the California fishery, causes great seasonal fluctuations as there is not sufficient reserve of fish to even out these variations, which are inevitable in biological populations.

CHAPTER 22

THE FISHERY NOW AND IN THE FUTURE

The Fishery

The rapid growth of the South African and South West African pilchard industries since the end of the Second World War has been accompanied by an increase in the size of the boats used and in the equipment carried to increase their working efficiency.

Pilchard vessels built shortly after the War were about 45 ft in length, but the most recent pilchard vessels have been in the region of 70 ft in length. The material used in building the hull is also undergoing change so that now fibreglass or steel-hulled boats rather than wooden vessels are appearing. Large diesel engines are used to power them and normal ancillary equipment generally now includes a puretic power block as well as radio telephone, a radio direction finder and echo-sounding apparatus. The vessels frequently have to search over a wider range than in the early days of the industry. In South West Africa most of the catch can still be obtained within 10 miles of the shore but the situation in South Africa is now rather different. In 1966 the South African Fish Meal Producers' Association (a body which with its sister organization, the S.A. Fish Oil P.A., deals with the marketing of the meal and oil from both South and South West Africa) chartered a 72-ft vessel to search an area between Hout Bay and Lambert's Bay up to 60 miles offshore for pilchard shoals. This vessel would then report back to any pilchard-catching vessel in the area when suitable shoals had been found (*South African Shipping News and Fishery Industry Review*, 1966d). This method has been superseded by the use of an aeroplane to assist in the spotting of shoals. The plane, costing £30,000, was purchased jointly by the pilchard factories of South Africa and was to operate at night guiding ships to the shoals which would be visible because of their phosphorescence (*South African Shipping News and Fishing Industry Review*, 1967h). The aeroplane was purchased at the end of the 1967 season and is not yet proven.

In the South and South West African pilchard fishery two active methods of fishing have been used. These are the lampara and purse-seine methods. Both types of net are encircling nets, but the lampara does not have the purse action which prevents the escape of fish from the bottom of the net. However, a lampara with a purse action has been developed in South Africa and this is known as the lampara seine net or the pursed lampara (von Brandt, 1964; du Plessis, 1958).

The main points of difference between the two types of net are that:

(i) the lampara has a dustpan-like action as the lower rope, the weighted ground rope, is much shorter than the floated upper rope, whereas the purse-seine has both upper and lower ropes of equal length;

(ii) the meshes along the length of the lampara vary: the rings are of coarse material and with larger meshes than the bag or bunt of the net, but in the purse-seine the meshes are of uniform size along the whole length and depth of the net;

(iii) the purse-seine has rings fitted to the lead line with a rope known as the purse line running through it. It is by this means that the net is closed at its base preventing the fish escaping by that route. The lampara does not possess this feature.

The South African pursed lampara retains the typical features of the lampara in that the cork line and lead line are of different lengths (170 fathoms and 145 fathoms, respectively). However, the meshes are the same throughout the length and depth of the net ($1\frac{1}{2}$ in. bar measurement), so that in this respect it resembles a purse-seine.

Furthermore, the pursing is effected in a similar way to that of the purse-seine with the purse lines passing through rings attached to the lead line. The pursed lampara, however, has two ropes attached to the deepest portion of the bag, known as the tongue, and one passing along the back of each ring.

Two boats are involved in the shooting of the net, one end being attached to a one-man dinghy, the other end usually being made fast to the mast of the parent vessel. The rowing boat encircles the shoal, passes a line to the main vessel after which both purse lines (or hauling lines) are winched in. The man in the dinghy has, meanwhile, attached his end of the net to floating drums and the dinghy to the cork line at the mid-line of the bag, and this prevents the net sinking under a heavy load of fish as it is pulled in.

During hauling, the area of the circle made by the net gradually decreases and becomes elongated in shape. When the tongue is brought alongside, the lines have been hauled in far enough for the lead lines to be retrieved and the wings and outer parts or shoulders of the net can then be hauled. This leaves the bag and the inner part of the shoulders containing the fish. The fish are removed, either by mechanical brailing or suction pumps, to the hold. The hauling process has recently been revised because, in the newest vessels, the puretic power block has been installed enabling the handling of the net to bring the lead lines aboard to be dispensed with.

The increasing size of the vessels and greater sophistication of the equipment used on board has led the price of the vessels to increase greatly during the past 15–20 years. For example, in the early 1950's a fishing boat about 50 ft long would cost about R25,000 with engine. The boats of about 70 ft length considered necessary today, plus all the equipment cost about R90,000, but more if the hull is made of modern materials (Trebett, 1967). As well as the equipment mentioned earlier this would also include a fish pump, which is increasingly being fitted, to draw the fish from the net into the hold.

The rising costs of vessels and equipment has made it more difficult for inshore fishing boats to be privately owned. Most boats are, therefore, owned by the factory although there is still a small percentage of privately owned vessels. In South Africa, of approximately 130 registered pilchard boats, nearly one-third is privately owned. The private owners, nevertheless, are licenced to fish for a certain factory, and each boat is allocated a percentage of the factory's quota. This worked out at between 3000 and 5000 short tons in 1962, depending on the size of the boat. It will be greater now owing to the increased factory quotas and the increasing hold capacity of the individual vessels. In 1966, the price paid to the fishermen was raised by 50 cents per ton bringing it to R9·30 per ton. At the time of this increase the fishermen were hoping to receive R10·0 per ton for the following reasons:

(i) the world price of fish meal had doubled since the last price on first landing was fixed, and it was thought that some of this profit should be passed on to the primary producers;
(ii) the cost of boats had doubled since the last increase;

(iii) the cost of maintenance had also doubled; and
(iv) the cost of living had increased considerably since the last increase in price of raw fish (*South African Shipping News and Fishing Industry Review*, 1966a).

The yield from the sales of the raw fish are distributed between the nine or ten crew members on a share basis which was operating in 1962 as shown below:

the skipper	received about R1 for each ton landed;
the engine-man	received about 65 cents for each ton landed;
the wheelman/mate	received about 65 cents for each ton landed;
and the ordinary crew man	received about 45 cents for each ton landed;

(*South African Shipping News and Fishing Industry Review*, 1962a).

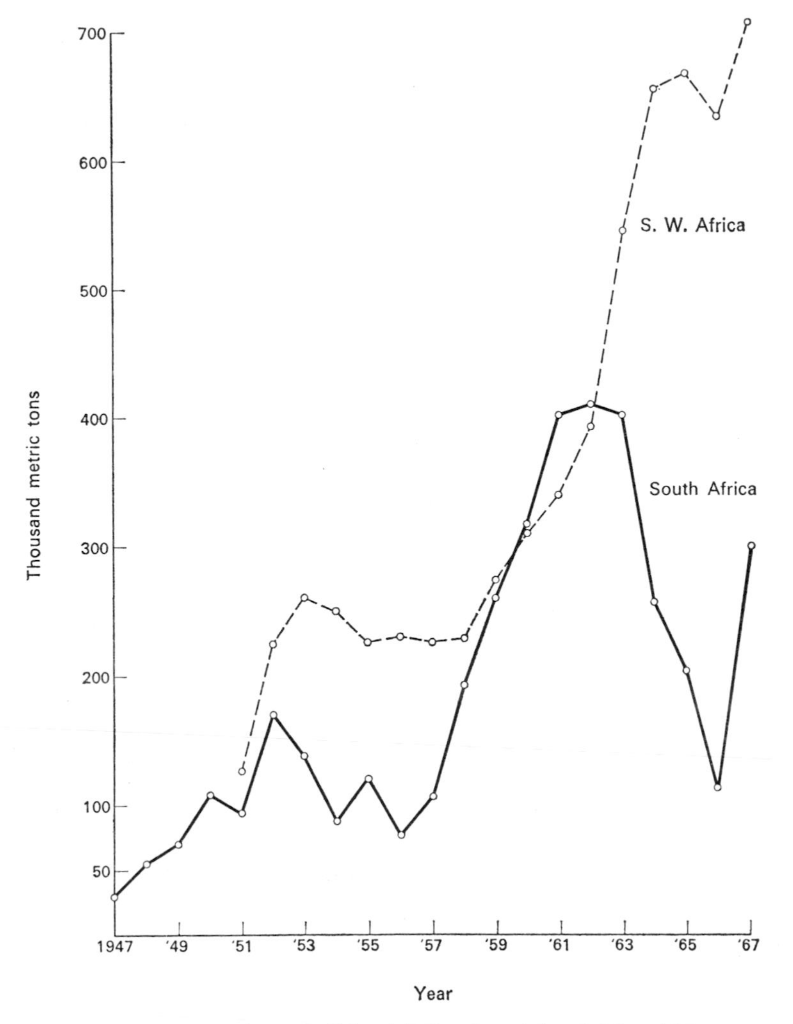

FIG. 22.1. Landings of pilchard in South and South West Africa.

In South Africa most of the fishing for, and processing of, pilchards is situated around St. Helena Bay where there are six factories connected with canning and reduction of pilchards. A little further south at Stompness Bay and Saldanha Bay there are three other factories while at Lambert's Bay to the north there are two more factories. In South West Africa, Walvis Bay is the main area of pilchard activity and there are nine concerns with factories there which are in some way connected with processing pilchards. At Lüderitz, to the south of Walvis Bay, there is also one reduction factory for pilchards, and in 1968 a reduction licence was granted to a firm which plans to set up its reduction plant at Cape Fria which is north of Walvis Bay (*South African Shipping News and Fishing Industry Review*, 1968c). This is partly as a result of the report of the Commission of Enquiry into the Fishing

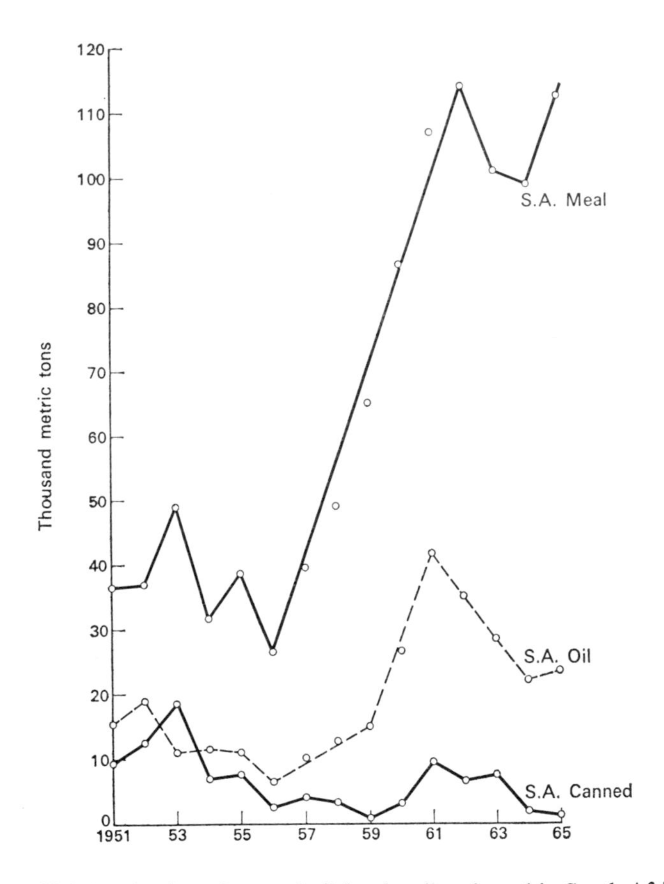

FIG. 22.2. Production of canned pilchards, oil and meal in South Africa.

Industry of South West Africa of which more will be said later in the chapter. Whilst this factory is being built its quota will be used by certain other factories and it is planned that part of the profits will be channelled into developing a harbour at Cape Fria.

Enough has already been said concerning the landings of pilchards in South and South West Africa to make further analysis unnecessary. They have been tabulated (Table 38) and have also been included here in graphical form (Fig. 22.1) to give a reminder of the pattern of events in the two countries.

An important influence on the landings of pilchards in both South and South West Africa has been the world demand for their products. The production of canned pilchards and of oil and meal are shown in graphical form (Figs. 22.2 and 22.3) and in the histograms (Figs. 22.4 and 22.5) for both countries separately and as a graph for the countries combined (Fig. 22.6).

Canned Pilchards

From the graph it can be seen that production has fluctuated considerably. These fluctuations have, of course, been caused by the demand for the product, largely from countries outside South Africa. South West Africa has, since 1954, invariably produced more canned pilchards than has South Africa. The United Kingdom has been placing regular and sizeable orders for African pilchards for many years, and in recent years the Philippines have ordered large quantities. To give an indication of the magnitude of exports to the United Kingdom the following two figures may be quoted (*South African Shipping News and*

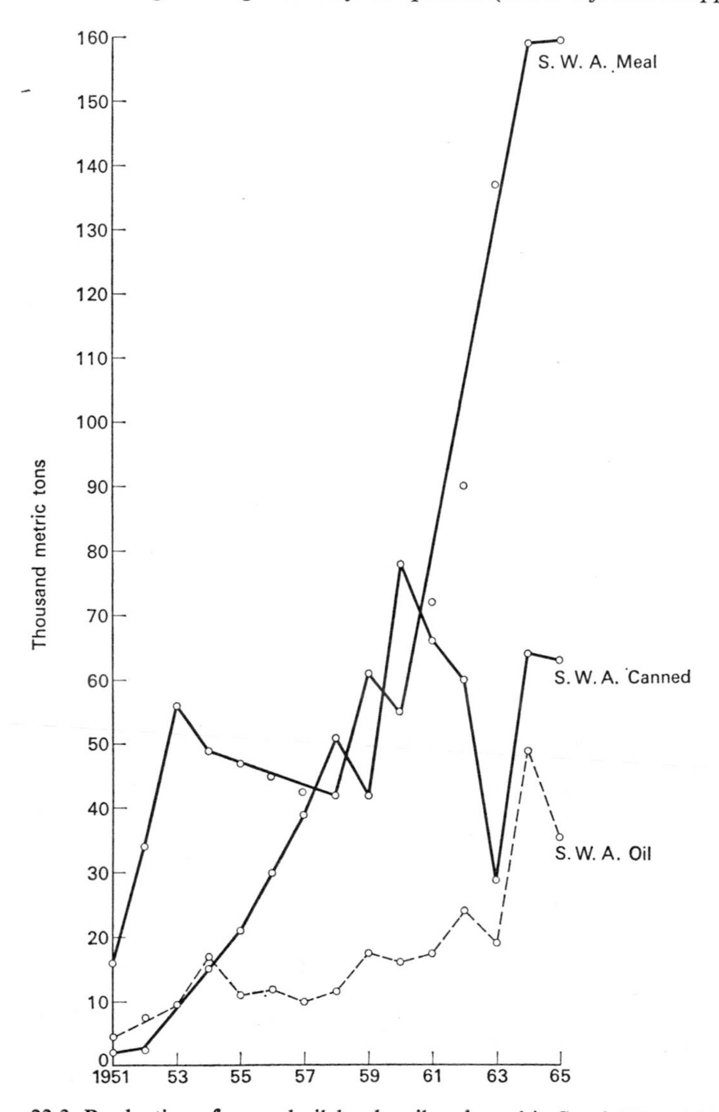

FIG. 22.3. Production of canned pilchards, oil and meal in South West Africa.

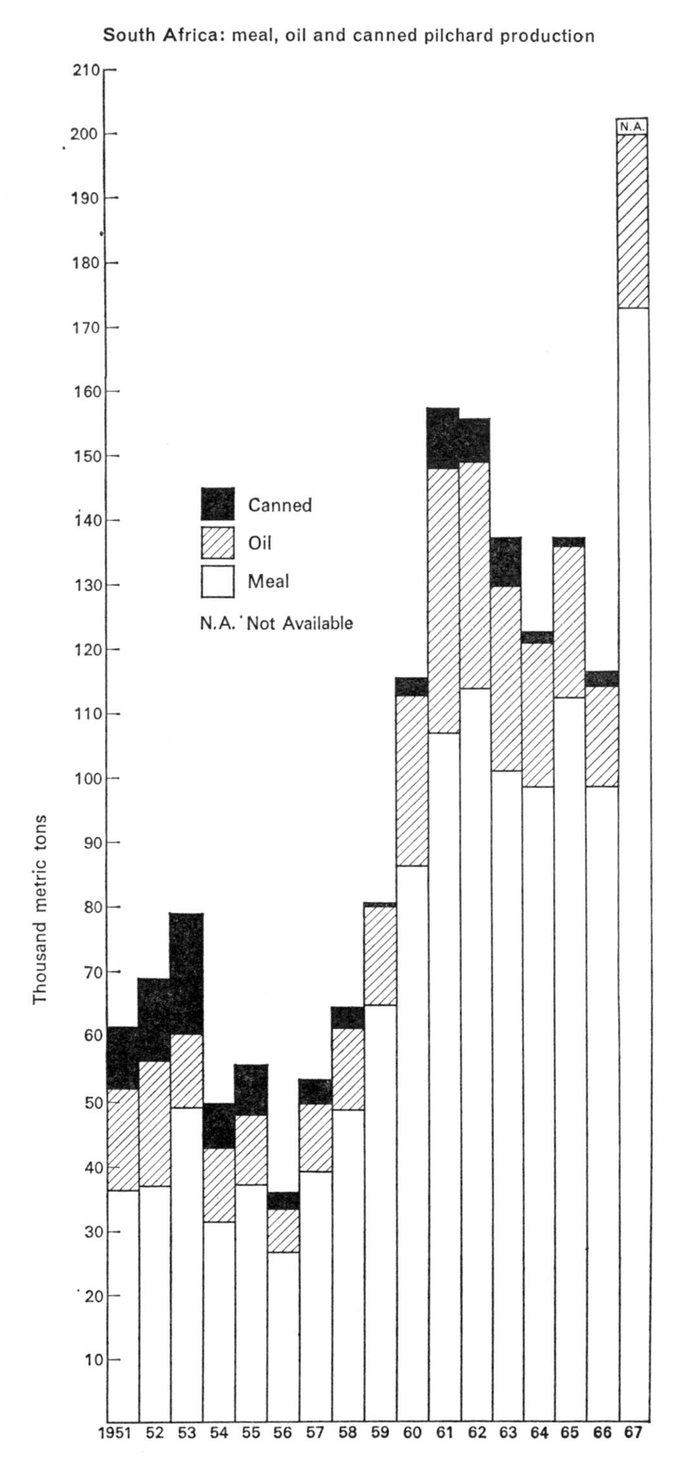

FIG. 22.4. Histograms of production in South Africa.

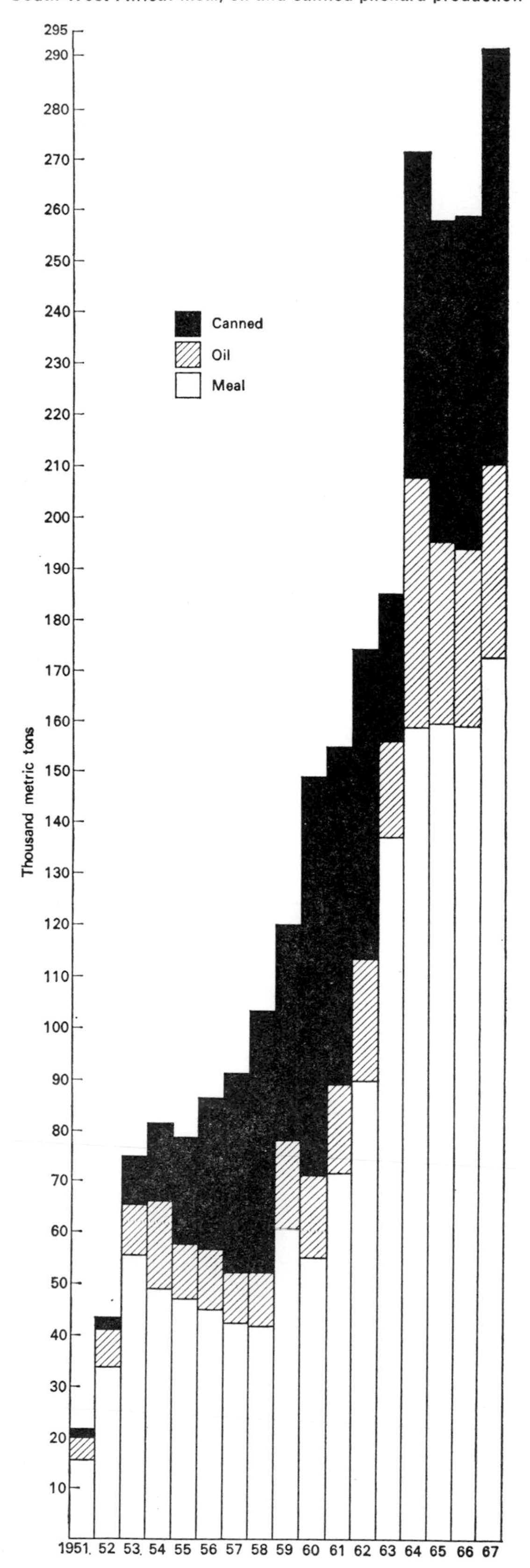

FIG. 22.5. Histograms of production in South West Africa.

Fishing Industry Review, 1967c). In 1965 exports totalled 776,866 cartons and in 1966, 1,034,800. That is, 16,919 and 22,536 tons respectively, a carton containing 48 lb of pilchards. In 1964 the order from the Philippines was for 1,375,000 cartons of canned pilchards (29,946 tons) negotiated with the Government Buying Agency of the Philippines. The Chairman of one of the South African fishing companies further reported that the "general world demand" required that there should be substantial canned fish production in 1965 (*South African Shipping News and Fishing Industry Review*, 1965b).

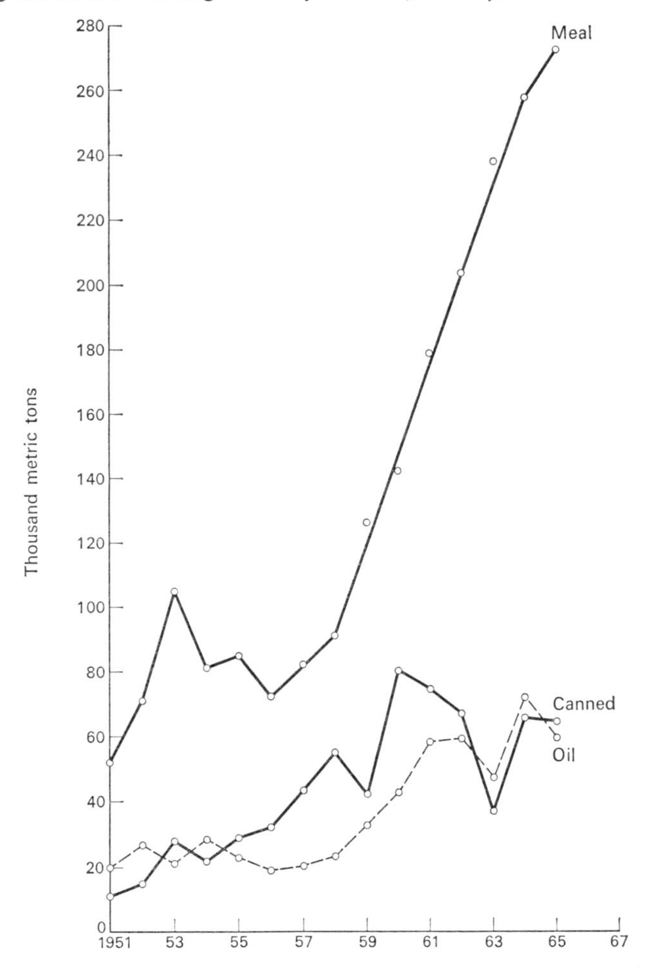

FIG. 22.6. Combined production of South Africa and South West Africa.

In these two industries there has not been the emphasis given to processing fish for human consumption that was so important at the beginning of the sardine industry in California. This is probably because a greater world demand for fish meal existed at the inception of these two industries than in 1914. Nevertheless, ready markets have been found for the canned product, and even on occasion some difficulty experienced in meeting the demand from some quarters (*South African Shipping News and Fishing Industry Review*, 1964e). The two figures quoted above for exports to the United Kingdom show that the product has not fallen from favour in that country in spite of some political difficulties. This is probably because the price has remained very low, in spite of a small increase in 1967 amounting to one-eighth of a penny on a 1-lb can, giving very good protein value for the money spent. The low price may previously have caused the product to be regarded as rather inferior, but it is possible (as suggested in *South African Shipping News and Fishing Industry Review*, 1967c) that recent economic difficulties in Britain have been advantageous to South African and South West African pilchard sales.

Compared with the production of fish meal, however, canned pilchards represent a small fraction (just under one-quarter in 1965, as shown in Fig. 22.6) of the total output of the two countries. A 1963 report to shareholders of one fishing concern (quoted in *South African Shipping News and Fishing Industry Review*, 1964c) stated that "there seems to be no doubt that, as long as certain bulk markets are closed to us, the overall production of canned fish from the factories will be somewhat limited". Production of canned pilchards was certainly exceptionally low in 1963 due, probably, to what was reported as "international fluctuations". (White Book on the activities of the South West African Administration for 1963 as quoted in *South African Shipping News and Fishing Industry Review*, 1964b.) However, the 1964 and 1965 figures showed an increase from this low figure. In 1966 there was an increase in both home and export sales for 1965 which was of the order of about 20 per cent in both spheres (*South African Shipping News and Fishing Industry Review*, 1967b) and this trend was expected to continue through 1967.

The most consistent buyers of canned pilchards from South and South West Africa have been the United Kingdom, the Philippines, the United States of America and Western Europe. It is probably for political reasons that other areas do not purchase these products and for which the "certain bulk markets" referred to above are closed. If political pressure should be lessened in the future then it is possible that canned pilchard production in the two countries could rise considerably. There have also been clear indications recently that their sales on the domestic market are increasing. For example, it has been stated (*South African Shipping News and Fishing Industry Review*, 1965b) that before 1964 home sales were at a static level of about 600,000 cartons of canned pilchards per year. In 1965, however, sales of this product were expected to reach 1,200,000 cartons internally (a forecast figure which was, in fact, exceeded), and the home sales forecast for 1966 was 1,750,000 cartons. This increasing demand is probably due to two factors, namely the increased income of the Bantu workers in the country and the low price of the product. Canned pilchards is one of the few protein products in South and South West Africa which has remained cheap; between 1948 and 1965 there was no price increase (*South African Shipping News and Fishing Industry Review*, 1966b).

Pilchard Oil and Meal

Production of oil and meal are linked processes and a further product known as stick-water also results. This is water removed from oil and meal during the processing and it contains many valuable constituents which are fed back into the meal at some stage of production.

Looking at the graph showing the combined production of meal for the two countries it can be seen that from 1956 through 1965 there was a continuous and rapid increase in the quantity of fish meal manufactured. Output in 1956 was about 72,000 tons, but by 1965 this had increased to 272,000 tons. However, the combined figures conceal a fall in the South African output in 1963 and 1964 followed by an almost complete recovery in 1965, to the

1962 level. This fall in production of pilchard meal from South Africa was masked in the overall figures by the very large increase in South West Africa which occurred in 1963. This was from 90,000 tons in 1962 to 137,000 tons in the following year: more than a 50 per cent increase. The decrease of meal production in South Africa coincides with the beginning of the decline in landings and a slight increase in production of canned pilchards.

In 1965 the increase in meal production in South Africa is probably accounted for by the increasing percentage of fish other than pilchards reduced to meal and included in the official statistics. For South Africa most of the annual totals for meal and oil production given in the official statistics include certain amounts of maasbanker and/or anchovy which have been reduced, but in the past this has been insignificant. With the decline of pilchard availability these other species have been sought with more interest and it is likely that this tendency will increase in the future. It will probably be the case that the anchovy industry in South Africa will have its own statistics shortly as it replaces the pilchard in importance in the shoal fish industry; the anchovy is inferior to the pilchard as a commercial proposition in several ways, as will be mentioned later.

Considering the production of fish meal between 1951 and 1956 the combined output (Fig. 22.6) rose from 1951 through 1953 and these showed a downward trend through until the end of 1956. It is possible that Peru's advent onto the world fish meal market in 1951 with a steadily increasing output, from 7300 tons in 1951 to 31,000 tons in 1956, caused the initial phase of the industries in South and South West Africa to be relatively depressed. World prices of fish meal have probably had some effect on general output of the product in that from one year to the next they have held firm or increased, thereby providing an atmosphere which was conducive to investment. A marked fall during 1960 (Peruvian meal was fetching $160 per ton in December 1959 but the price fell during the year to $60 per ton in September 1960) could possibly account for the fall in the South West African output for that year. A report in *South African Shipping News and Fishing Industry Review*, (1960) stated that prices in mid-1960 were from 30 to 50 per cent lower than in mid-1959.

Production of pilchard oil in the two countries is interesting in that, although it follows similar trends to those of fish meal production, as would be expected, there have been more fluctuations within the main trends. This is the result of the oil content of the fish varying, as it does considerably, from season to season as well as from month to month within seasons. For example, Fig. 22.6 shows a marked fall in the output of oil for 1963 in spite of a continued increase in meal production. This was because the oil yield during 1963 was only from 8 to 10 gal/ton in 1963 compared with 25–28 gal/ton in 1964 (*South African Shipping News and Fishing Industry Review*, 1964d) and an average of about 13–14 gal/ton. Hjul

TABLE 46. AVERAGE MONTHLY OIL YIELDS PER TON OF SOUTH AFRICAN PILCHARDS, 1957 THROUGH 1961

Month	Yield (gal/ton)	Month	Yield (gal/ton)
April	13	August	12
May	16	September	11
June	18	October	8
July	17		

(1962) gave the average oil yields for the normal 7 fishing months over the 5-year period from 1957 through 1961. These are shown in Table 46.

Variations in oil yield also occur at the different fishing ports. In 1965, pilchards processed at Walvis Bay gave 14 gal/ton whereas those at Lüderitz yielded only 6 gal/ton (*South African Shipping News and Fishing Industry Review*, 1966c).

The fall in production of fish oil in 1963 seemed to have the effect of increasing the price of oil. It was mentioned earlier that no difficulties have really been experienced in selling the products of the pilchard industry and the strong demand for the oil during the 1963 season when it was in short supply caused the prices at the end of that season to close more than 100 per cent higher than at the beginning of the season (*South African Shipping News and Fishing Industry Review*, 1964a).

Throughout the history of the two industries there have been continual efforts made to improve the efficiency of the processing plants involved in the manufacture of all three products.Therefore, as well as the increase in throughput of the factories, the rise in output is also partly due to technical improvements which have occurred. In 1961 it was recorded (*South African Shipping News and Fishing Industry Review*, 1961c) that the 1961 catch was less than double that of 1958, but more than double the fish meal and three times the quantity of fish body oil were obtained.

One of the technical improvements which came into general use in 1962–3 was concerned with the transfer of offal from the cannery to the fish meal plants. Before this time it was transferred by fluming. This is a process whereby the material was washed down chutes to the cannery holding tank. Now, however, the offal is fed onto conveyor belts of Neoprene which transfer the material in a dry state to the cannery. This has eliminated the leaching of some oil and particulate matter which occurred in the former process and as a result the yields of meal and oil have been improved.

Other products are also manufactured by some factories processing pilchards. For example, a South African factory produces canned curried pilchards and canned smoked pilchards in oil. In the latter product the pilchards are smoked for 60 to 70 min and then canned with an addition of sild oil (*South African Shipping News and Fishing Industry Review*, 1962b). Pilchards are also frozen for bait and for human consumption. One white fish factory at Walvis Bay has a licence to process a quantity (10,000 short tons) of pilchards for these purposes. Seventy per cent of the production is for human consumption. For this, they are processed at the rate of 2 tons each hour being frozen to —28°F, held at that temperature for 16 hr and then held in cold store at —20°F. When required they are packed in thin blocks weighing 5 lb and dispatched in cardboard units of 50 lb. That intended for bait is also frozen to —28°F and held in a cold store at —20°F but is then packed in large 40-lb blocks into cardboard containers (*South African Shipping News and Fishing Industry Review*, 1966g). In 1967 a factory more directly connected with the pilchard industry at Walvis Bay also started freezing pilchards. This firm uses horizontal plate freezers and produces 10-lb units which are similarly dispatched in cartons of 50 lb. These are then stored in a 250-ton cold store until required. This product is eaten by the African workers in the mines and is transported inland by rail in special trucks in which they are packed in flake ice (*South African Shipping News and Fishing Industry Review*, 1968b). The same outlet is also used for some pilchard cutlets frozen at another Walvis Bay factory, but these also apparently find markets in the United Kingdom and Germany. This factory also freezes pilchard in the round, and these are used for bait by fishing vessels of South African and other nationalities (*South African Shipping News and Fishing Industry Review*, 1968c). These are comparatively recent developments and it remains to be seen what importance they may assume in the future. However, once the difficulties in transporting a frozen product have been overcome, as they seem to have been here, it is usually found that the convenience of these products makes them popular. With a healthy internal demand coming from the mining districts of South Africa it would seem that this product might assume considerable significance in the future, even if more difficulties are experienced in extending markets in other parts of the world. A final specialized product is also worthy of mention and that is the canning of pilchards in soya sauce for the Japanese market. What quantities are concerned in these various products is not shown by the statistics but it seems that the South and South West African producers are keen to fulfil any demands for unusual preparations, and statistics on these would certainly be of interest.

The awareness of the industry in both these countries to general world conditions of demand for their products (the marketing side of both countries works through common organizations) is shown, for example, in the production of the specialized products described above; the willingness to try new gear to aid in the catching side as exemplified in the wide acceptance of the Puretic power block and the fish pump; the way in which the needs of the fishermen are anticipated by the shore-based management as shown by the purchase of the aeroplane to aid in spotting shoals now that they are becoming more difficult to find; and finally the frequent modification of processing plant as new developments make it possible to improve the end product, all show that the pilchard industries of South and South West Africa are vital and highly organized right through from the location of the fish shoals to the final distribution of the product.

In fact, the overall organization of these two inshore fisheries is such that it compares in several important ways with the organization of the middle-distance and distant-water trawler fleet of the United Kingdom. For example, the considerable number of factories involved in Walvis Bay and in the Cape are controlled by comparatively few individuals. The directors of one company are frequently found to be directors of several companies. Furthermore, as already mentioned, it is the processing companies which own the majority of fishing boats of the industry so that much greater control and co-ordination over catching and processing projects can be exercised.

On the production side there are three organizations, the South African Fish Canners' Association Ltd., the South African Fish Meal Producers' Association Ltd., and the South African Fish Oil Producers' Association Ltd., whose declared aims are the co-ordination and secretarial organization of the three main processing branches of the industry. In addition, the whole of the marketing of pilchard meal and pilchard oil is organized by the two relevant associations and at least part of the marketing of the canned product is arranged by the S.A. Fish Canners' Association.

Here again it is found that the board of directors of the three associations contain a large degree of overlap and are also involved in other companies with producing interests. Therefore, it can be seen that a large amount of vertical integration as well as considerable horizontal integration has occurred in the industry. This has undoubtedly enabled the rapid expansion of the industry to take place as it has been possible to control all aspects of production without one becoming out of phase with another. If some similar state of organization had arisen in the Cornish pilchard industry even on a much smaller scale, the fatal state of disorganization reached there in the 1950's would not have occurred. The large development of the canning side of the industry in Cornwall made the industry top-heavy and, without comparable expansion on the catching side, was bound to result in failure. It is possible that in the long term some of the canning firms in Cornwall could have remained solvent and even prospered if they had been prepared to invest in the catching side of the industry on the one hand and in the marketing of the product on the other.

The emphasis on research and carefully considered development which was shown in the South and South West African pilchard industries during their early days has recently been revived in the Commission of Inquiry appointed by the South West African Administration to investigate all aspects of the South West African fishing industry. Their terms of reference as quoted in *South African Shipping News and Fishing Industry Review* (1966i) were:

> To enquire into and report and make recommendations on the systematic purpose and effective exploitation of and control over the fish potential of South West Africa, both with regard to the current quotas already granted to industrialists and the question of the desirability of increasing or decreasing such quotas or of granting one or more new licenses either in Walvis Bay or elsewhere along the coast of South West Africa.
>
> The following specific matters have also been referred to the Committee for investigation:
> (i) the production of fish meal flour for human consumption;
> (ii) applications for a quota of pilchards for freezing and marketing;
> (iii) the building of a quay at Walvis Bay;
> (iv) the application for a quota of pilchards by the white fish industry.

The report of the Committee was published in March 1967 and the summary of the recommendations was very fully reported in *South African Shipping News and Fishing Industry Review* (1967i). Much emphasis was placed on the undesirability of the South African factory ships and it was specifically stated that attempts should be made to halt their activities. One interesting approach was the apparent threat that should there appear to be any signs of depletion of the pilchard resource which could be attributable to the activities of the South African factory ships, then "South West Africa should apply its whole existing potential as speedily as possible for exploitation". More is said below concerning the factory ship controversy. There is a continued healthy attitude to fishery research in the report. A fisheries research board should, it was recommended, be instituted which would control the fisheries research laboratory. This would be responsible for the administration and financing of the laboratory and it would create a technical advisory committee which would be responsible for the determination of research projects and their priorities. It would also supervise the projects and consider where and how the results could be most advantageously applied to the industry. An example of the care with which new projects are approached was illustrated on the question of fish flour. No permit for a manufacturing plant should be given for this product without "quantitative proof . . . of the marketing and profitability of the flour". Further details of the report can be seen in the reference already cited, but it remains to be seen how closely the Administration of South West Africa will apply the recommendations.

There are, however, in South and South West Africa several aspects which have developed recently which poise question marks over the future and continued health of pilchard fishing there. The first of these has already been mentioned, and that is, in South Africa itself, the rapid decline of total landings of pilchards in recent years. It may be, as has been suggested by some authorities (for example, Trebett, 1967) that this decline is part of a rather long-term cyclical pattern not yet fully understood. The alternative opinion has already been expressed in this book that there has been overfishing of the pilchard stocks of the Cape in spite of early conservation attempts.

The pelagic fish landings in South Africa have, however, been maintained at a high overall figure as anchovy has been sought to replace the pilchard. These two species obviously occupy very similar ecological niches and it seems that when competition is left in a natural balance the pilchard dominates. However, when this balance is severely and constantly upset by heavy fishing mortality of the pilchard then the anchovy begins to compete successfully, presumably in the early larval stages, with the pilchard.

This view seems to be shared by de Jager (see, for example, in Trebett, 1967) and, as he points out, the increase in fishing effort on the anchovy stocks may help to redress the balance in favour of the pilchard once more. It is hoped, by those with commercial interests in the fishing industry in South Africa, as well as by others with a purely humanitarian approach to food production, that the pilchard soon returns to its former prominence in South African waters. This is because it is superior to the anchovy in four ways. First, at present the anchovy is not canned in South Africa, treating it for human consumption being "both time-consuming and difficult" (Report of the Fisheries Development Corporation of South Africa Limited for 1965–6 as given in *South African Shipping News and Fishing Industry Review*, 1967a). Next, the oil yield per ton is less than that from pilchards; thirdly, it spoils more readily, and finally it is more expensive to equip a boat to catch anchovy, an anchovy net costing about R10,000 compared with about R5000 for a pilchard net.

In view of the difficulties which the industry is experiencing in South Africa the policy of the administrators who are empowered to increase the licenced fishing effort becomes increasingly hard to understand. For example, in 1965 two new fish-meal licences were granted at the Cape. This was at the time when the previous season's pilchard catch had been the lowest since 1958 and had continued in a very marked way the decline started in 1963. The sharpness of the fall in landings is well shown in Fig. 22.1, it being even more marked coming, as it does, after the continuous increase from 1956 through 1962. It would seem a logical step for the licencers to have taken to ban all further increases in fishing power in 1964 until they could see whether the fall in landings was to be halted quickly or not. As stated in the previous chapter, the administration had abandoned the quota system and here in 1964, with all the signs of overfishing confronting them, two further licences for meal manufacture were granted. It was estimated that for these to be used to full advantage between twenty and thirty shoal boats would have to be added to the fleet (*South African Shipping News and Fishing Industry Review*, 1965a). Not all the increased effort represented by these licence increases would be expended solely on catching pilchards as anchovy and maasbanker are also shoal fish which are useful fish for reduction purposes, but for the reasons pointed out above the pilchard is the most sought after for reduction and would be taken whenever possible. One estimate of the increase of available fishing effort in the South African fleet as a result of the two new licences was that the overall fleet strength would be increased by up to 20 per cent (*South African Shipping News and Fishing Industry Review*, 1966e).

Another development, referred to above, which has recently occurred in the South African pilchard industry, and which is bound to have a marked effect on the future of the pilchard stocks in those waters as well as in the waters off South West Africa, is the licensing by the Government of three factory ships to fish for pilchards. These factory ships are not allowed to fish within the 12-mile territorial limits of South or South West Africa. In fact, by 1968 only two, the *Willem Barendsz* and *Suiderkruis*, were operating. The *Willem Barendsz* was the first factory ship to be owned by a South African company and several of the faults in design, which, for example, meant that time spent in port between fishing trips was greater than desirable, were eliminated in the conversion of the *Suiderkruis*. Conversion of these huge vessels was extremely costly and details of them can be found in *South African Shipping News and Fishing Industry Review* (1966f and 1967e), but the large throughput available in their operation must make them attractive commercial propositions. The processing plant of the *Suiderkruis*, for example, can deal with 1680 short tons of wet fish per day.

The large quantity of fish necessary to satisfy the processing equipment is caught by a fleet of seiners, nine in the case of the *Willem Barendsz* and eighteen for the *Suiderkruis*. Experience with the former's fleet showed that larger seiners were better than conventional sized vessels. This may mean that new pilchard vessels could be over a hundred feet in length as the factory ship was using one vessel 110 ft long. It has been estimated (*South African Shipping News and Fishing Industry Review*, 1968a) that the total quantity of pilchards taken by the two factory ships in 1968 will be in the region of 350,000 short tons (about 318,000 metric tons), their total catch from January to the end of September in 1967 was 248,880 short tons but not both vessels were in operation for the whole season (*South African Shipping News and Fishing Industry Review*, 1968d).

At present there are only two large South African factory ships, but as other companies see the success with which these two operate they will almost certainly take steps to capitalize on similar schemes. There are already two smaller factory–purse-seiners being built in Norway to operate in South African or South West African waters and these will be based in ports of that area. The developer of this idea envisages a fleet of ten such vessels (*South African Shipping News and Fishing Industry Review*, 1967f), and this is only the third year of factory-ship philosophy in Southern Africa.

So that it seems, that by licencing the operation of factory ships from their shores, the South African Government has unleashed a tremendous new force on to the fishing scene. This development has not been looked upon at all favourably by the Administration in South West Africa. They view these ships, and possibly correctly so, as a potential hazard operating just outside their territorial limits. (They are not allowed to fish within the 12-mile limit of either country.) Initially the ships were not allowed to enter South West African ports, but the latest official attitude seems to be that if they wish to take on provisions or have repairs done at South West African ports then a duty of 50 per cent is to be levied on all goods supplied to them (*South African Shipping News and Fishing Industry Review*, 1967d). The same source states that if they are found to be fishing within South West African territorial waters then the vessel and gear and its catch can all be confiscated and a fine of R10,000 is also payable by the owners. To enforce the fishing limits, South West Africa was reported to be buying some patrol boats.

When the licences were first granted the Chairman of the Committee of Enquiry into the Fishing Industry of South West Africa attempted to have them annulled and approached the South African Prime Minister and the Minister of Economic Affairs. He pointed out that the two countries had agreed not to issue further licences without prior joint consultation and he, therefore, assumed that the factory vessels would not operate north of the Orange River; that is, they would not even enter the extra-territorial waters off South West Africa. (Reported in *South African Shipping News and Fishing Industry Review*, 1966h.)

It is possible to consider the advent of the factory ships, as one chairman of a South African fishing company has recently done, as "a natural development, in that it opens up to further exploitation the rich shoal fish resources off the South African coast" (*South African Shipping News and Fishing Industry Review*, 1967a). On the other hand, it is just this so-called opening-up process which is so worrying to the Administration in South West

Africa. The pilchard shoals off South West Africa for the 5-year period 1963 through 1967 yielded over 500,000 tons of fish annually and the possibility of this yield being increased by more than half as much again by the factory ships is a serious development which should not be allowed without much research into the strength of the resource. As the chairman of the South African fishing company quoted above also stated: "the matter of conservation and protection becomes more important and also more urgent. These are problems for which no ready solutions offer themselves." It would seem that for the continued health of the resource these are the very words which should make all concerned with the licencing of factory ships call a halt to this possibly disastrous method of exploitation.

It should be admitted by commercial concerns and government bodies as well as by the *South African Shipping News and Fishing Industry Review* (1966h) that "the factory ships are primarily a means of circumventing the established conservation measures of fish processing licences, catching capacity limits and raw fish quotas which are enforced in South and South West Africa". The above journal then states that if these vessels are allowed to fish then it would be logical "to give the shore-based pilchard industry a bigger share in the total catch". This would prevent the fishing boats and processing plants being idle for 3 months of the year. This is true if these vessels are officially recognized. However, it would appear that as these factory ships are evading legal restrictions imposed for the good of the fishery, then official approval should be withdrawn and other regulations concerning the operation of their supply fleets be tightened up. All this should be done before too many concerns are involved and before any more capital is invested in this form of fishing.

Possibly the greatest fact in favour of the South African policy is that the waters off South and South West Africa have now become internationally famous for their richness. There are almost continually some foreign fishing vessels, including factory ships from Russia, operating in these waters. It is difficult, therefore, to preach restraint to the South African and South West African fishing concerns when other countries are sweeping up great quantities of fish only a few miles from their shores. It is in this area of the world that the most recent need for international co-operation in fishery resource conservation has arisen. It is a problem which requires immediate attention and one which, with all the research that has been applied to the industry, should be capable of resolution.†

It would seem that the first step that will have to be taken is for a conference to be called of all countries which have interests in fishing for pilchards off South and South West Africa. From such a conference possibly some sort of international quota system, based on the existing value of investment in the industry or some other suitable criterion or criteria, would result. This would enable the fishery to be pursued without any country feeling that too much of an attitude of first-come, first-served, was necessary in the utilization of this remarkable resource. It is sincerely to be hoped that some international policy of exploitation can be forged before the resource itself demonstrates once again that the overfishing of a pelagic species, at least economically, is a possibility to be reckoned with.

† A recent article by Gulland and Carroz deals with this problem and mentions some ways in which international agreement on utilization of fishery resources might be reached. The article, entitled "Management of Fishery Resources", appears in Russell and Yonge (1968).

PART V

STATISTICAL APPENDIX

CHAPTER 23

APPENDIX OF STATISTICS OF SOME OTHER SARDINE/PILCHARD INDUSTRIES OF THE WORLD

In this brief section the figures of annual landings given in the F.A.O. *Yearbooks of Fishery Statistics* have been used. Those countries which, over the 7-year period 1961 through 1967, landed an average of 10,000 metric tons or more of sardine or pilchard have been listed. (The landings of the United Kingdom, South Africa, South West Africa and California are also included again for completeness and for ease of comparison.) The figures have all been taken from F.A.O. 1968a (Table C 2-5 of that volume). Former editions of the *Yearbooks* showed a separate entry for the Philippines but in the F.A.O. 1968a the landings of that country have been placed under the heading of "various marine clupeoids" so that it is not certain to what species this may be referring, but it is likely that the landings of sardines in the Philippines are considerable. (Average approximately 30,000 metric tons p.a.)

Two points of particular interest emerge from this world survey of the catches of sardines and pilchards. The first concerns the European pilchard. Table 47 shows that there are six countries which land over 10,000 tons p.a. of the European pilchard, *Sardina pilchardus*. It can be seen that not all the figures are available, but the mean total landings for the European pilchard for these six countries (Algeria, Morocco, France, Italy, Portugal and Spain) from 1961 through 1967 is 470,000 tons. The landings for Greece are only available for 3 years and so have not been included in the table, but figures for those years show that catches in that country are over 10,000 tons p.a. (For 1964, 1965 and 1966 landings were 13·0, 10·6 and 11·4 thousand tons respectively.) Of the six countries France, possibly, and Morocco, definitely, are increasing their landings of *Sardina pilchardus* and the other four are remaining remarkably static. (Chile and India are increasing their landings of *Sardinops sagax* and *Sardinella longiceps* respectively.)

Riedel (1960) has written an interesting survey of "Sardine production off the Atlantic coast of Europe and Morocco since 1920" and gives tables and graphs illustrating the relative importance of the sardine or pilchard catch in England, France, Spain and Portugal. Comparatively little is known in detail about the resource of these or of the Mediterranean countries, although tagging experiments have been carried out in some countries (e.g. Yugoslavia, see Mužinić, 1960a) and attempts to determine the composition of the stock and the catch per unit effort have also been made (e.g. in Spain, see Mužinić, 1960b). Although *Sardina pilchardus* is found from the eastern Mediterranean, round the Atlantic coast of Europe and into the English Channel it is not known how continuous or discontinuous this stock is. However, certain tagging experiments and other circumstantial evidence suggest that, in the Mediterranean Sea at least, there are several distinct stocks which undergo only very limited migrations (Mužinić, 1960c). Perhaps a corollary to this, certain morphological

TABLE 47. COUNTRIES LANDING OVER 10,000 TONS/YEAR SARDINE OR PILCHARD. CATCH FROM 1961 THROUGH 1967

('000 metric tons)

	1961	1962	1963	1964	1965	1966	1967	1968
AFRICA								
Algeria[c]	14·6	n.a.	9·2	10·4	10·6	13·0	n.a.	
Angola[e]	55·6	77·3	75·9	108·4	56·8	69·8	56·2	
Morocco[c]	123·9	125·6	127·2	139·7	160·1	252·0	208·1	
South Africa[e]	402·4	410·7	400·9	256·1	204·1	114·2	299·2	
South West Africa[e]	342·3	394·7	546·7	655·9	666·0	634·7	707·4	
NORTH AMERICA								
California[a,g]	19·6	7·0	3·2	6·0	0·9	0·4	†	
Mexico[a]	20·3	14·8	19·3	19·0	20·3	18·6	29·4	
SOUTH AMERICA								
Chile[b]	26·5	24·6	27·9	37·4	43·0	62·1	122·4	
ASIA								
India[f]	167·9	110·3	63·6	274·3	262·0	252·8	259·1	
Japan[d]	127·1	108·0	55·9	16·2	9·2	13·5	16·0	
EUROPE								
France[c]	28·0	37·3	32·2	31·2	24·3	38·7	46·8	
Italy[c]	41·6	28·7	29·9	27·1	29·6	31·6	26·9	
Portugal[c]	139·4	130·5	118·6	163·6	138·0	125·0	n.a.	
Spain[c]	140·9	109·3	124·0	119·4	113·7	119·8	105·6	
United Kingdom[c,g]	2·7	2·0	2·0	1·7	1·4	1·4	0·7	
Soviet Union	4·1	6·1	29·2	79·3	111·0	52·4	41·6	

[a] *Sardinops caerulea.*
[b] *Sardinops sagax.*
[c] *Sardina pilchardus.*
[d] *Sardinops melanosticta.*
[e] *Sardinops ocellata.*
[f] *Sardinella longiceps.*
[g] Included for comparison purposes.
† Negligible quantity.
n.a. Not available.

differences have been noted, in particular an increase in the vertebral number from south to north (Furnestin, 1952). A useful summary of the biology of the sardine of the Mediterranean is to be found in Larreñatá (1960).

In the Mediterranean Sea the species is exploited as sardines and this also applies to Moroccan, Spanish and Portuguese fisheries of the Atlantic coasts. France takes both adult and young forms of the species. The different ages at which exploitation occurs are linked with the method of fishing. Those fisheries using purse-seines or some other type of encircling net usually catch sardines, whereas drift-net fisheries seem to be gauged to catch the older fish. In the Mediterranean there is frequent use of powerful lights to attract the fish to the nets and in some parts of France bait is used to concentrate the fish before being encircled (known as the *sardine de rogue*).

It can be seen that much basic work remains to be done on this important resource; not only in England where the lack of fundamental data has already been made obvious, but also in all the other countries catching it. The statistics show that in certain areas the landings are increasing annually and it is likely that further modernization of the fishing industry in other countries (e.g. Yugoslavia and Greece) will enable their landings to be increased further. It would be advantageous to the countries concerned to organize a joint research programme so that advance biological knowledge of the stock would enable them to utilize their resource to the best advantage. Increasing sophistication within their fishing industries will bring with it associated problems of fair exploitation, for example, those connected with fishery limits. Co-operation in the scientific field should facilitate the solving of the international fisheries problems which will possibly arise in that area quite soon.

The second point of interest concerns the Russian landings of *Sardinops ocellata*. These can be seen to have risen from practically nothing in 1961 to over 100,000 tons in 1965. In 1966 and 1967 there was a decline again but it is likely that this is only temporary. Russian factory vessels are frequently reported in *South African Shipping News and Fishing Industry Review* to be fishing off the coasts of South and South West Africa. There are also reports of the building or converting of vessels for this purpose. Here is a concrete example of the sort of difficulty referred to above. As was pointed out in the last chapter this exploitation of a country's fish stocks by another country can lead to wrong decisions being taken concerning fishery management. It was shown how the fishing of factory vessels has been allowed by South Africa in a way which contravenes earlier conservation attempts. This

TABLE 48. UTILIZATION: PRODUCTION OF DRIED SARDINE OR PILCHARD
('000 metric tons)

	1961	1962	1963	1964	1965	1966	1967	1968
AFRICA								
Algeria		n.a.						
Angola			3·8[a]	4·5[a]	5·0	n.a.	n.a.	
Morocco		0·3	0·5	n.a.	0·2	n.a.	n.a.	
South Africa								
South West Africa								
NORTH AMERICA								
California		†	0·1	†	†	†	n.a.	
Mexico		0·1	†	†	†	0·4	1·0	
SOUTH AMERICA								
Chile		†	0·1	†				
ASIA								
India								
Japan	80·3	83·5	77·9	74·0	72·0	57·9	60·5	
EUROPE								
France	3·5	2·0[a]	2·6[a]	3·1[a]	n.a.	n.a.	n.a.	
Italy	26·0	3·2[a]	3·0[a]	3·0[a]	2·7[a]	2·7[a]	3·0[a]	
Portugal								
Spain		n.a.	†					
United Kingdom		†	n.a.	n.a.	n.a.	n.a.	n.a.	

† Negligible quantity.
n.a. Not available.
[a] Species not specified.

has probably been encouraged or justified by the fact of foreign vessels operating near South Africa's territorial waters. As was stressed in the last chapter, some international policy is needed, in cases such as this area off South and South West Africa and in the Mediterranean Sea, so that the fishery resource can be properly managed and yet reasonably distributed among the nations using it.

The utilization of the catch is shown in Tables 48–52. Details are not always as clear as would be ideally desired, but the overall pattern of production can be seen from them. The source of information is F.A.O. (1968b) for figures from 1962 through 1967 (Tables C1-4, E1-2, G1-5, H1-3) and F.A.O. (1962) for the 1961 figures (Tables F9, F16, F26, F30). There are no remarkable trends other than those which have already been discussed for South Africa and South West Africa, except that the tremendous increase of fish meal production by Russia is shown. It is not known how much of this is pilchard meal as the statistics are not specific enough but it is an interesting development. There are no figures for production of canned pilchards by Russia. It is known that the Russian factory ships operating off South Africa do manufacture this product but what percentage of their catch is directed into oil, meal or canned product is not given.

TABLE 49. UTILIZATION: PRODUCTION OF SARDINE OR PILCHARD WHICH HAS BEEN SALTED AND DRIED,[a] SALT-PRESSED,[b] WET-SALTED[c] OR PRESERVED IN BRINE[d]

('000 metric tons)

	1961	1962	1963	1964	1965	1966	1967	1968
AFRICA								
Algeria								
Angola								
Morocco								
South Africa								
South West Africa								
NORTH AMERICA								
California								
Mexico								
SOUTH AMERICA								
Chile								
ASIA								
India								
Japan	33·8[a] 4·1[c]	26·8[a] 4·5[c]	27·1[a] 3·1[c]	23·5[a] 2·5[c]	24·5[a] 2·4[c]	29·6[a] 2·8[c]	26·8[a] 4·2[c]	
EUROPE								
France								
Italy								
Portugal	0·4[d]	0·2[d]	0·4[d]	0·8[d]	0·5[d]	0·6[d]	0·7[d]	
Spain		4·4[b] 0·2[d]	4·7[b] 0·1[d]	4·6[b] 0·1[d]	5·6[b] 0·1[d]	4·5[b] 0·1[d]	4·3[b] 0·1[d]	
United Kingdom								

TABLE 50. UTILIZATION: PRODUCTION OF CANNED PILCHARD OR SARDINE
('000 metric tons)

	1961	1962	1963	1964	1965	1966	1967	1968
AFRICA								
Algeria		n.a.	0·2[a]					
Angola								
Morocco		36·2	30·5	41·9	43·8	42·0	43·7	
South Africa	9·4	6·7	7·7	1·8	1·3	2·3	n.a.	
South West Africa	65·7	60·5	29·1	56·4	63·2	65·4	81·9	
NORTH AMERICA								
California	8·6	2·8	1·2	2·5	0·2	0·1	†	
Mexico	8·3	7·3	10·6	10·5	11·2	11·9	19·1	
SOUTH AMERICA								
Chile	2·3	1·9	2·7	2·3	2·5	2·7	4·0	
ASIA								
India								
Japan	12·7	12·4	4·5	4·6	3·8	5·4	3·6	
EUROPE								
France	26·5	28·6	19·0	21·0	18·0	28·6	n.a.	
Italy	4·5	4·3	1·6	2·4	2·5	2·5	2·5	
Portugal	54·8	54·6	49·6	70·2	56·2	52·4	48·8	
Spain	21·3	15·4	17·3	16·9	18·9	25·1[a]	22·2	
United Kingdom	†	†	2·7[a]	2·7[a]	2·4[a]	—	—	

† Negligible quantity.
n.a. Not available.
[a] Species not specified.

TABLE 51. UTILIZATION: PRODUCTION OF OIL FROM PILCHARD OR SARDINE UNLESS OTHERWISE NOTED
('000 metric tons)

	1961	1962	1963	1964	1965	1966	1967	1968
AFRICA								
Algeria								
Angola	3·5[a]	3·8[a]	3·6[a]	7·4[a]	5·6[a]	4·6[a]	4·8[a]	
Morocco			5·2[a]	5·3[a]	5·4[a]	6·0[a]	5·6[a]	
South Africa	41·1[b]	35·1[b]	28·6[d]	22·2[b]	23·3[b]	15·6[b]	27·0[b]	
South West Africa	17·4	26·7	18·9	47·2	35·4	34·5	37·6	
NORTH AMERICA								
California	0·3	0·1	1·6[c]	1·3[c]	1·7[c]	2·4[c]	2·6[c]	
Mexico	0·6	0·2	0·4[a]	0·5[a]	0·5[a]	0·3[a]	0·4[a]	
SOUTH AMERICA								
Chile	3·9[c]	12·1[a]	12·5[a]	17·6[a]	10·4[a]	22·3[a]	14·6[a]	
ASIA								
India		0·1[a]	†	0·1[a,e]	0·5[a,e]	†[a]	0·1[a]	
Japan	0·9	0·8	0·6	0·1	0·4	0·5	0·2	
EUROPE								
France								
Italy								
Portugal		3·6[a]	3·6[a]	5·4[a]	5·4[a]	4·4[a]	4·7[a]	
Spain		n.a.	n.a.	n.a.	n.a.	n.a.	0·6	
United Kingdom								

† Negligible quantity.
n.a. Not available.
[a] Species not specified.
[b] Includes maasbanker.
[c] Includes other fish.
[d] Includes maasbanker and anchovy.
[e] Based on export data.

TABLE 52. UTILIZATION: PRODUCTION OF MEAL FROM PILCHARD OR SARDINE UNLESS OTHERWISE NOTED
('000 metric tons)

	1961	1962	1963	1964	1965	1966	1967	1968
AFRICA								
Algeria								
Angola	55·3[a]	33·0[a]	32·8[a]	54·7[a]	46·4[a]	48·1[a]	40·4[a]	
Morocco		16·0[a]	19·9[a]	21·8[a]	22·0[a]	29·0[a]	35·0[a]	
South Africa	107·0[b]	113·9[b]	101·1[b]	98·7[b]	112·6[b]	98·6[b]	172·6[b]	
South West Africa	71·9	89·6	137·1	153·8	155·2	158·8	172·1	
NORTH AMERICA								
California	2·3	0·6	n.a.	n.a.	n.a.	n.a.	n.a.	
Mexico	6·7[c]	4·5[c]	5·3[c]	5·5[c]	7·1[c]	9·9[c]	10·2[c]	
SOUTH AMERICA								
Chile	3·8	3·3	3·5	5·4	6·3	9·6	20·1	
ASIA								
India								
Japan			n.a.	n.a.	n.a.	n.a.	n.a.	
EUROPE								
France	15·5[a]	10·0[a]	11·0[a]	12·0[a]	13·2[a]	13·6[a]	13·2[a]	
Italy		n.a.	n.a.	n.a.	n.a.	2·0[a]	2·0[a]	
Portugal	7·4[a,c]	5·4[a,c]	4·7[a,c]	7·4[a,c]	8·0[a]	10·9[a]	7·0[a]	
Spain	n.a.	n.a.	n.a.	n.a.	n.a.	n.a.	4·3	
United Kingdom								
Soviet Union	100·6[a,c]	91·5[a,c]	112·7[a,c]	144·7[a,c]	202·6[a,c]	238·5[a,c]	294·9[a,c]	

n.a. Not available.
[a] Species not specified.
[b] Includes maasbanker.
[c] May include some meal of white fish origin.

BIBLIOGRAPHY

Note. Where possible abbreviations for references have been taken from BROWN, PETER and STRATTON, G. B. (Eds.) (1965) *World List of Scientific Periodicals 1900–1960*, Butterworths, London.

AGASSIZ, A. (1883) Exploration of the surface fauna of the Gulf Stream III: The Porpitidae and Velellidae, *Mem. Mus. comp. Zool. Harvard* **8** (2).

AHLSTROM, E. H. (1943) Studies on the Pacific pilchard or sardine, 4. Influence of temperature on the rate of development of pilchard eggs in nature, *Spec. scient. Rep. U.S. Fish Wildl. Serv.* **23,** 132–67.

AHLSTROM, E. H. (1954) Distribution and abundance of egg and larval populations of the Pacific sardine, *Fishery Bull. Fish and Wildl. Serv. U.S.* **93** (56), 81–140.

AHLSTROM, E. H. (1960) Synopsis on the biology of the Pacific sardine (*Sardinops caerulea*), *Proceedings of the World Scientific Meeting on the Biology of Sardines and Related Species.* II—Species synopses, subject synopses, 415–51, F.A.O., Rome.

ARTHUR, D. K. (1956) The particulate food and food resources of the larvae of three pelagic fishes, especially the Pacific sardine, *Sardinops caerulea* (Girard), Ph.D. thesis, University of California, Scripps Institution of Oceanography.

BAINBRIDGE, RICHARD (1953) Studies on the interrelationships of zooplankton and phytoplankton, *J. Mar. Biol. Ass. U.K.* **32** (3), 385–448.

BAKER, E. A. and FOSKETT, D. J. (1958) *Bibliography of Food*, Butterworths Scientific Publications.

BENHAM, FREDERIC (1955) *Economics. A General Introduction*, Pitmans, London, p. 394.

BENNETT, W. J. (1952) A century of change on the coast of Cornwall, *Geography* **37,** 214–24.

BLAXTER, J. H. S. and HOLLIDAY, F. G. T. (1963) The behaviour and physiology of herring and other clupeids, in RUSSELL, F. S. (Ed.) *Advances in Marine Biology* **I,** 261–393, Academic Press, London and New York.

BÖHNECKE, GÜNTER and DIETRICH, GÜNTER (1951) Monatskarten der Oberflächentemperatur für die Nord- und Ostsee und die augrendzenden Gewässer, *Deutsches Hydrographisches Institut*, Hamburg.

BRIDGER, J. P. (1965) The Cornish pilchard and its fishery, *Laboratory Leaflet* (new series), vol. 9 Min. Ag. Fish. Food, Lowestoft.

BROEK, VAN DEN C. J. H. (1965) Fish canning, in BORGSTROM, GEORG (Ed.) *Fish as Food*, vol. 4. *Processing*, vol. 2. Academic Press, New York and London.

BUYS, M. E. L. (1959) The South African pilchard (*Sardinops ocellata*) and maasbanker (*Trachurus trachurus*), Hydrographical environment and the commercial catches, 1950–57, *Div. Fish. Invest. Rep.* **37,** 176 pp., Pretoria.

CALIFORNIA (1950) *Prog. Rep. Calif. Coop. Sardine Res. Progm.*, State of California, Marine Research Committee.

CALIFORNIA (1952) *Prog. Rep. Calif. Coop. Ocean. Fish. Invest.*, 1 Jan. 1951–30 June 1952, 51 pp.

CALIFORNIA (1965) The California marine fish catch for 1963, *Fish Bull. Calif.* **129.**

CALIFORNIA (1967) The California marine fish catch for 1965, *Fish Bull. Calif.* **135.**

CALIFORNIA (1968) The California marine fish catch for 1966, *Fish Bull. Calif.* **138.**

CAREW, R. (n.d.) *Survey of Cornwall*, This was originally in the late sixteenth and early seventeenth centuries. It was later edited by TONKIN, THOMAS, and published by Lord de Dunstanville in 1811.

CHU, S. P. (1946) The utilization of organic phosphorus by phytoplankton, *J. Mar. Biol. Ass. U.K.* **26** (3), 285–95.

CLARK, F. N. (1934) Maturity of the California sardine (*Sardinops caerulea*) determined by ova diameter measurements, *Fish Bull. Calif.* **42,** 49 pp.

CLARK, F. N. (1939) Can the supply of sardines be maintained in California waters? *Calif. Fish Game* **25** (2), 172–6.

CLARK, F. N. (1940) The application of sardine life-history to the industry, *Calif. Fish Game* **26** (1), 39–48.

CLARK, F. N. (1952) Review of the California sardine fishery, *Calif. Fish Game* **38** (3), 367–80.

CLARK, F. N. and JANSSEN, J. F. (1945a) Movements and abundance of the sardine as measured by tag returns, *Fish Bull. Calif.* **61,** 7–42.

CLARK, F. N. and JANSSEN, J. F. (1945b) Measurement of the losses in the recovery of sardine tags, *Fish Bull. Calif.* **61,** 63–90.

CLARK, F. N. and MARR, J. C. (1955) Population dynamics of the Pacific sardine, *Prog. Rep. Calif. Coop. Ocean. Fish Invest.*, 1 July 1953–31 March 1955, pp. 11–48.

Clowes, A. J. (1950) An introduction to the hydrology of South African waters, *Div. Fish. Invest. Rep.* **12,** Pretoria.

Colman, J. S. (1950) *The Sea and Its Mysteries*, Bell, London, 285 pp.

Cooper, C. E. B. (1957) Chilling of pilchards, *Fish. Ind. Res. Inst.*, 9th Annual Report, Cape Town.

Cooper, L. H. N. (1937) On the ratio of nitrogen to phosphorus in the sea, *J. Mar. Biol. Ass. U.K.* **22** (1), 177–82.

Cornish, Thomas (1884) Mackerel and pilchard fisheries, in *Conferences Held in Connection with the Great International Fisheries Exhibition*, 1883, vol. III, pp. 109–46. William Clowes, London.

Couch, Jonathon (1835) Essay on the natural history of the pilchard, *Royal Cornwall Polytechnic Society*, 3rd Annual Rep., 65–101.

Couch, Jonathon (1869) *A History of the Fishes of the British Islands*, vol. IV, Groombridge, London.

Crawford, Sir William and Broadley, H. (1938) *The People's Food*, Heinemann, London and Toronto, 336 pp.

Crutchfield, James and Zellner, Arnold (1963) Economic aspects of the pacific halibut fishery, *Fishery Industrial Research*, **1** (1), Washington.

Culley, M. B. (1965) Pilchard fisheries: a study, *Fishg. News Int.* **4** (2), 183–8.

Cunningham, J. T. (1891) The reproduction and growth of the pilchard, *J. Mar. Biol. Ass. U.K.* **2** (2), 151.

Cushing, D. H. (1953) Studies on plankton populations, *J. Cons. Perm. Int. Explor. Mer.* **19** (1), 1–22.

Cushing, D. H. (1957) The number of pilchards in the Channel, *Fish. Invest.* Ser. II **21** (5), H.M.S.O., London.

Cushing, D. H. and Harden Jones, F. R. (1968) Why do fish school? *Nature* **218** (5145), 918–20.

Cutting, C. L. (1955) *Fish Saving. A History of Fish Processing from Ancient to Modern Times*, Leonard Hill (Books) Ltd.

Daugherty, A. E. and Wolf, R. S. (1960) Age and length composition of the sardine catch off the Paciffc coast of the United States and Mexico in 1957–58, *Calif. Fish Game* **46** (2), 189–94.

Daugherty, A. E. and Wolf, R. S. (1964) Age and length composition of the sardine catch off the Pacific coast of the United States and Mexico in 1961–62, *Calif. Fish Game* **49** (4), 241–52.

Davies, D. H. (1956a) The South African pilchard (*Sardinops ocellata*). Sexual maturity and reproduction, 1950–54, *Div. Fish. Invest. Rep.* **22,** 1–55, Pretoria.

Davies, D. H. (1956b) The South African pilchard (*Sardinops ocellata*). Migration, 1950–55, *Div. Fish. Invest. Rep.* **24,** Pretoria.

Davies, D. H. (1957) The South African pilchard. Preliminary report on the feeding of the pilchard off the West Coast, 1953–56. *Div. Fish. Invest. Rep.* **30,** Pretoria.

Davies, D. H. (1957a) The biology of the South African pilchard, *Div. Fish. Invest. Rep.* **32,** Pretoria.

Davies, D. H. (1958a) The South African pilchard (*Sardinops ocellata*) and maasbanker (*Trachurus trachurus*). The predation of sea-birds in the commercial fishery, *Div. Fish. Invest. Rep.* **31,** Pretoria.

Davies, D. H. (1958b) The South African pilchard (*Sardinops ocellata*). Preliminary report on the age composition of the commercial catches, 1950–55, *Div. Fish. Invest. Rep.* **33,** Pretoria.

Davis, F. M. (1937) An account of the fishing gear of England and Wales. *Fish Invest.* Ser. II **XV** (2).

Davis, F. M. (1958) An account of the fishing gear of England and Wales. *Fish Invest.* Ser. II **XXI** (8), 165 pp. H.M.S.O., London.

Davis, H. C., Clark, G. H. and Shaw, P. A. (1945) Chilling sardines. Experience with refrigeration as an aid to canning, *Pacific Fisherman*, April, pp. 37, 39.

Day, Francis (1880) *The Fishes of Great Britain and Ireland*, vol. 2, pp. 224–30, London.

Dewberry, E. B. (1954) British Columbia's salmon-canning industry, *Fd. Process. Packag.* **23,** 250–5.

Dewberry, E. B. (1957) Canning pilchards in Cornwall, *Fd. Manufacture* **32** (6), 268.

Dietrich, Günter (1951) Die anomale Jahresschwankung des Wärmeinhalts im Englischen Kanal, ihre Ursachen und Auswirkungen, *Deutsches Hydrographisches Institut, Ozeanographie* 1950, pp. 184–201.

Dreosti, G. M., Cooper, C. E. B., Rowan, A. N., Rowan, M. K. and Nachenius, R. J. (1957) Mush in canned pilchards, *Fish Ind. Res. Inst., Progr. Rep.* **22,** 14 pp., Cape Town.

Duhamel du Monceau, H. L. (1772) *Traité général des pěches, et historie des poissons*, Paris, **2,** sec. 3, pp. 420–56.

Fage, Louis (1920) La sardine (*Clupea pilchardus* Walb.), *Report on the Danish Oceanographical Expeditions, 1908–10, to the Mediterranean and Adjacent Seas*, vol. II, Biology, A.9 Engraulidae, Cludeidae, pp. 34–93.

Felin, F. E. (1954) Population heterogeneity in the Pacific pilchard, *Fish Bull. Calif.* **54** (86), 201–25.

Felin, E. F., Daugherty, A. E. and Pinkas, Leo (1950) Age and length composition of the sardine catch off the Pacific coast of the United States in 1949–50, *Calif. Fish Game* **36** (3), 241–9.

Felin, E. F., Daugherty, A. E. and Pinkas, Leo (1951) Age and length composition of the sardine catch off the Pacific coast of the United States in 1950–51, *Calif. Fish Game* **37** (3), 339–49.

Felin, E. F., Anas, Ray, Daugherty, A. E. and Pinkas, Leo (1952) Age and length composition of the sardine catch off the Pacific coast of the United States in 1951–52, *Calif. Fish Game* **38** (3), 427–35.

Fishing News (1962) 26 January, No. 2358, 3, London.

Fishing News (1962a) 22 June, No. 2559, 8, London.
Fishing News (1962b) 7 December, No. 2583, 2, London.
Fishing News (1965) 17 December, No. 2741, 8, London.
FISHING NEWS (BOOKS) LTD. (1964) *Modern Fishing Gear of the World* **2,** pp. 352–4, London.
FLEMING, R. H. (1939) The control of diatom populations by grazing, *J. Cons. Perm. Int. Explor. Mer.* **14** (2), 210.
FOOD AND AGRICULTURAL ORGANIZATION OF THE UNITED NATIONS (1955) *Yearbook of Fishery Statistics,* 1952–3, **IV** (1), F.A.O., Rome.
—— (1958) *Report of the Technical Meeting on Costs and Earnings of Fishing Enterprises,* The point of view of government and other public authorities. Min. Ag. Fish. Fd. and The Scottish Home Dept., U.K., 34, Rome.
—— (1960) *Yearbook of Fishery Statistics* **II,** 1959, F.A.O., Rome.
—— (1962) *Yearbook of Fishery Statistics* **14,** 1961, F.A.O., Rome.
—— (1965) *Yearbook of Fishery Statistics* **18,** 1964, F.A.O., Rome.
—— (1968a) *Yearbook of Fishery Statistics* **24,** *Catches and Landings 1967,* F.A.O., Rome.
—— (1968b) *Yearbook of Fishery Statistics* **25,** *Fishery Commodities 1967,* F.A.O., Rome.
FOSBERG, F. R. (1963) The island ecosystem, in FOSBERG, F. R. (Ed.) *Man's Place in the Island Ecosystem, A Symposium* **1–6,** Honolulu.
FRAMPTON, R. H. C. F. (1954) *An Account of the Methods of Fishing, and Part iculars of Gear used in the South-west,* Min. Ag. Fish. Fd., H.M.S.O., London.
FURNESTIN, J. (1952) Biologie des clupéidés méditerranéens. Oceanographie méditerranéene, *Vie et Milieu,* suppl. 2.
GATES, D. E. (1961) The Pacific sardine—*Sardinops caerulea, Calif. Ocean Fish. Resources to the Year 1960,* pp. 46–48.
GIRARD, CHARLES (1854) Descriptions of new fishes, collected by Dr. A. L. Heerman, naturalist attached to the survey of the Pacific railroad route, under Lieut. R. S. Williamson, U.S.A., *Acad. Nat. Sci. Philadelphia,* Proc. 7, 138.
GOODRICH, E. S. (ED.) (1930) *A Treatise on Zoology,* Part IX—*Cyclostomes and Fishes,* Cambridge Univ. Press.
GOSLINE, W. A. (1960) Contribution toward a classification of modern isospondylous fishes, *Bull. Brit. Mus. (Nat. Hist.) Zool.* **6,** 327–65.
GRAHAM, MICHAEL (1943) *The Fish Gate,* 111, London.
GREAT BRITAIN (1945–66) *Sea Fisheries Statistical Tables,* Min. Ag. Fish. Fd., H.M.S.O., London.
GREAT BRITAIN (1949–56) *Annual Report and Accounts,* White Fish Authority, H.M.S.O., London.
GREAT BRITAIN (1954) *Annual Report and Accounts,* year ended 31 March 1954. White Fish Authority, H.M.S.O., London.
GREAT BRITAIN (1957–61) *Domestic Food Consumption and Expenditure,* Annual reports of the National Food Survey Committee, H.M.S.O., London.
GREAT BRITAIN (1958) *Annual Report and Accounts,* for the year ended 31 March 1958, White Fish Authority, H.M.S.O., London.
GREAT BRITAIN (1959) *Annual Report and Accounts,* for the year ended 31 March 1959, White Fish Authority, H.M.S.O., London.
GREAT BRITAIN (1960) *Annual Report and Accounts,* for the year ended 31 March 1960, White Fish Authority, H.M.S.O., London.
GREAT BRITAIN (1960a) *Herring Industry Board,* 26th Annual Report, Cmnd. 1366, H.M.S.O., London.
GREAT BRITAIN (1960b) *Domestic Food Consumption and Expenditure,* Annual report of the National Food Survey Committee, 27–35, H.M.S.O., London.
GREAT BRITAIN (1961) *Report of the Committee of Inquiry into the Fishing Industry,* Cmnd. 1266, 84, H.M.S.O., London.
GREAT BRITAIN (1961a) White Fish Authority. Pilchard industry development project, *Progress Bulletin* 1, London.
GREAT BRITAIN (1961b) Herring Industry Board, 27th Annual Report, Cmnd. 1723, H.M.S.O., London.
GREAT BRITAIN (1962) White Fish Authority, information supplied by South-western Area Office, 7.11.62.
GREAT BRITAIN (1962a) Ministry of Agriculture, Fisheries and Food. Information supplied by London Office, 8.2.62.
GREAT BRITAIN (1963) Tariff and import policy division, notice to importers, No. 1035, Board of Trade, London.
GREAT BRITAIN (1963a) White Fish Authority, Information supplied by South-western Area Office, 17.6.63.
GREAT BRITAIN (1963b) White Fish Authority, Information supplied by South-western Area Office, 1963.
GREAT BRITAIN (1966) White Fish Authority, Information supplied by South-western Area Office, 8.11.66.
GREAT BRITAIN (1966a) Ministry of Agriculture, Fisheries and Food. Information supplied by London Office, 20.10.66.
GREENWOOD, P. H., ROSEN, D. E., WEITZMAN, S. H. and MYERS, G. S. (1966) Phyletic studies of Teleostean fishes, with a provisional classification of living forms, *Bull. Am. Mus. Nat. Hist.* **131,** (4).

HAND, C. H. and BERNER, LEO (1959) Food of the Pacific sardine (*Sardinops caerulea*), *Fishery Bull. Fish Wildl. Serv. U.S.* **60** (164), 175–84.

HARDY, A. C. (1936) Observation on the uneven distribution of the Oceanic plankton, *Discovery Rep.* **11,** 511–38.

HARDY, A. C. (1956) *The Open Sea. Its Natural History: I. The World of Plankton*, Collins, London, 335 pp.

HARDY, A. C. (1959) *The Open Sea. Its Natural History: II. Fish and Fisheries*, Collins, London.

HARDY, A. C. and GUNTHER, E. R. (1935) The plankton of the South Georgia whaling grounds and adjacent waters, 1926–27, *Discovery Rep.* **11,** 1–456.

HART, J. L. (1937) A brief account of the life-history of the pilchard, *Rep. Prov. Dep. Fish. Br. Columb.*, T 50–T 56.

HARVEY, H. W. (1925) Evaporation and temperature changes in the English Channel, *J. Mar. Biol. Ass. U.K.* **13** (3), 678–92.

HARVEY, H. W. (1930) Hydrography of the mouth of the English Channel 1925–1928, *J. Mar. Biol. Ass. U.K.* **16** (3), 791–820.

HARVEY, H. W. (1934) Annual variation of planktonic vegetation, 1933, *J. Mar. Biol. Ass. U.K.* **19** (3), 775–92.

HARVEY, H. W. (1950) On the production of living matter in the sea off Plymouth, *J. Mar. Biol. Ass. U.K.* **29** (1), 97–137.

HARVEY, W. H., COOPER, L. H. N., LEBOUR, M. V. and RUSSELL, F. S. (1935) Plankton production and its control, *J. Mar. Biol. Ass. U.K.* **20** (2), 407–40.

HATTON, S. R. and SMALLEY, G. R. (1938) Reduction processes for sardines in California, *Calif. Fish Game* **24** (4), 391–414.

HICKLING, C. F. (1939) The selective action of the drift-net on the Cornish pilchard, *J. du. Cons.* **XIV** (1), 67–80.

HICKLING, C. F. (1945) The seasonal cycle in the Cornish pilchard, *J. Mar. Biol. Ass. U.K.* **26** (2), 115–38.

HIGGINS, ELMER and HOLMES, H. B. (1921) Methods of sardine fishing in southern California, *Calif. Fish Game* **7** (4), 219–37.

HJORT, J. (1926) Fluctuations in the year-classes of important food fishes, *J. Cons. Perm. Int. Explor. Mer.* **I** (5).

HJUL, PETER (1962) Six factories processed a record pilchard catch, *S. Afr. Shipp. News and Fish. Ind. Rev.* **XVII** (1), 53–61.

HODGSON, W. C. (1957) *The Herring and its Fishery*, Routledge & Kegan Paul, London, 197 pp.

HODGSON, W. C. and RICHARDSON, I. D. (1949) The experiments on the Cornish pilchard fishery in 1947–1948, *Fish. Invest.* Ser. II, **17** (2), H.M.S.O., London.

HOLLIMAN, E. S. and OVENDEN, A. E. (1959) Concepts, definitions and conventions, in *Report of the Technical Meeting on Costs and Earnings of Fishing Enterprises*, London, 8–13 September 1958, vol. 66.

HOLME, N. A. (1961) The bottom fauna of the English Channel, *J. Mar. Biol. Ass. U.K.* **41** (2), 397–461.

HOLME, N. A. (1966) The bottom fauna of the English Channel, Part II, *J. Mar. Biol. Ass. U.K.* **46** (2), 401–93.

HUBBS, C. L. (1929) The generic relationships and nomenclature of the California sardine, *Calif. Acad. Sci. Proc.* **4,** Ser. **18** (11), 261–5.

HUGHES, R. B. (1957) The quality of canned herring, *British Fd. Manufacturers Ind. Res. Ass.* Tech. Circular 120.

IRVIN and JOHNSON (1963) *South African Fish and Fishing*, Irvin & Johnson, Cape Town.

JAGER, B. v. D., DE (1960) Fluctuations and biotic factors, *Proceedings of the World Scientific Meeting on the Biology of Sardines and Related Species*. I—Report, 7–14, F.A.O., Rome.

JAGER, B. v. D., DE (1960a) Synopsis on the biology of the South African pilchard *Sardinops ocellata* (Pappé), *Proceedings of the World Scientific Meeting on the Biology of Sardines and Related Species*, II—Species synopses, subject synopses, pp. 97–114, F.A.O., Rome.

JANSSEN, J. F. and APLIN, J. A. (1945) The effect of internal tags upon sardines, *Fish Bull. Calif.* **61,** 43–62.

JARVIS, N. D. (1950) Curing of fishery products, *Res. report* 18, *Fish Wldl. Serv. U.S.* 27 pp.

JENYNS (afterwards BLOMEFIELD), LEONARD (1842) Notes on a collection of fishes from San Diego, California, in JORDAN, D. S. and GILBERT, C. H. (1881) *Proc. U.S. Nat. Mus.* (for 1880) **3,** 30.

KENNEA, T. D. (1957) A geographical study of the fishing industry of the English south coast, M.Sc. thesis, University of London.

KERKUT, G. A. (1960) *Implications of Evolution*, Pergamon Press, Oxford.

KRISTJONSSON, HILMAR (Ed.) (1959) *Modern Fishing Gear of the World*, Fishing News (Books) Ltd., London, pp. 400–13.

KWAN-MING, LI (1960) Synopsis on the biology of *Sardinella* in the tropical eastern Indo-Pacific area, *Proceedings of the World Scientific Meeting on the Biology of Sardines and Related Species*, II—Species synopses, subject synopses, pp. 175–212, F.A.O., Rome.

LARREÑATÁ, M. G. *et al.* (1958) Captures por unidad de esfuerzo en la pesqueria de sardina de Castellon, *Invest. Pesq.* **12**, 49–81.

LARREÑATÁ, M. G. (1960) Synopsis of biological data on *Sardina pilchardus* of the Mediterranean and adjacent seas, *Proceedings of the World Scientific Meeting on the Biology of Sardines and Related Species*, II—Species synopses, subject synopses, pp. 137–73, F.A.O., Rome.

LEARSON, R. J. (1969) Breaded fishery products. In FIRTH, F. E. (Ed.), *The Encyclopedia of Marine Resources*, Van Nostrand Reinhold Co., New York, pp. 66–71, 740 pp.

LEBOUR, M. V. (1921a) The larval and post-larval stages of the pilchard, sprat and herring from the Plymouth district, *J. Mar. Biol. Ass. U.K.* **12** (3), 427–57.

LEBOUR, M. V. (1921b) The food of young clupeids, *J. Mar. Biol. Ass. U.K.* **12** (3), 458–67.

LEONARD HILL (BOOKS) (1957) *Food Industries Manual*, 18th ed., London, 1035 pp.

LUCAS, C. E. (1936) On certain inter-relations between phytoplankton and zooplankton under experimental conditions, *J. Cons. Int. Explor. Mer.* **11**, 343–62.

LUCAS, C. E. (1947) The ecological effects of external metabolites. *Biol. Rev.* **22**, 270–95.

LUMBY, J. R. (1925) A note on the water movements in the English Channel and neighbouring seas, considered on the basis of salinity, *J. Mar. Biol. Ass. U.K.* **13** (3), 670–72.

LUMBY, J. R. (1935) Salinity and temperature of the English Channel. Atlas of Charts, *Fish. Invest.* Ser. II, **14** (3), H.M.S.O., London.

MALCOLM-SMITH, E. F. (1949) *Encyclopaedia Britannica*, vol. 4, p. 748, London.

MARCH, E. J. (1955) Herring drifting in the days of sail—the ways of the Cornishmen, *World Fish.* **4** (9), 51–53.

MARCHAND, J. M. (1951) Pilchard research programme, *22nd Annual Report, Div. Fish.*, 15–21, Pretoria.

MARINE BIOLOGICAL ASSOCIATION OF THE U.K. (1957) *Plymouth Marine Fauna*, 3rd edn.

MARR, J. C. (1951) On the use of the terms abundance, availability and apparent abundance in fishery biology, *Copeia* 1951 (2), 163–9.

MARR, J. C. (1957) The sub population problem in the Pacific sardine, *Sardinops caerulea*, *Spec. Sci. Rep. U.S. Fish Wildl. Serv.* (208), pp. 108–25.

MARR, J. C. (Rapporteur) (1960) Report of Section I. Population identification. *Proceedings of the World Scientific Meeting on the Biology of Sardines and Related Species*, I—Report, 3–6, F.A.O., Rome.

MARR, J. C. (1960a) The causes of major variations in the catch of the Pacific sardine, *Sardinops caerulea*, *Proceedings of the World Scientific Meeting on the Biology of Sardines and Related Species*, III—Stock and Area Papers, pp. 667–791, F.A.O., Rome.

MARSHALL, N. B. (1965) *The Life of Fishes*, Weidenfeld & Nicolson, London.

MATTHEWS, J. (Rapporteur) (1960) Working group report IV. Sardine fishery statistics, *Proceedings of the World Scientific Meeting on the Biology of Sardines and Related Species*, I—Report, pp. 36–42, F.A.O., Rome.

MATTHEWS, J. P. (1690a) Synopsis on the biology of the South West African pilchard *Sardinops ocellota* (Pappé), *Proceedings of the World Scientific Meeting on the Biology of Sardines and Related Species*, II—Species synopses, subject synopses, pp. 115–35, F.A.O., Rome.

MATTHEWS, J. P. (1960b) The pilchard of South West Africa (*Sardinops ocellata*). Size composition of the commercial catches in South West Africa, 1952–57, *Mar. Res. Laboratory*, *Invest. Rep.* **1**, South West Africa.

MATTHEWS, J. P. (1961) The pilchard of South West Africa (*Sardinops ocellata*) and the maasbanker (*Trachurus trachurus*), Bird predators, 1957–1958, *Mar. Res. Laboratory*, *Invest. Rep.* **3**, South West Africa.

MATTHEWS, J. P. (n.d.) The pilchard of South West Africa (*Sardinops ocellata*). Sexual development, condition factor and reproduction, 1957–1960, *Mar. Res. Laboratory*, *Invest. Rep.* **10**, South West Africa.

MILLER, D. J. (1956) The anchovy, *Prog. Rep. Calif. Coop. Ocean. Fish Invest.* 1 April 1955–30 June 1956, pp. 20–27.

MORGAN, ROBERT (1956) *World Sea Fisheries*, Methuen, London.

MURPHY, G. I. (1961) Oceanography and variations in the Pacific sardine population, *Prog. Rep. Calif. Coop. Ocean. Fish Invest.* **VIII**, 55–64.

MURPHY, G. I. (1965) Population dynamics of the Pacific sardine, (*Sardinops caerulea*), Ph.D. thesis, University of California, San Diego.

MURPHY, G. I. (1966) Population biology of the Pacific sardine (*Sardinops caerulea*), *Proc. Calif. Acad. Sci.*, 4th Series, **XXXIV** (1), 84 pp.

MUŽINIČ, RADOSNA (1960a) *Sardina pilchardus* Walb.—Mediterranean and Black Sea, *Proceedings of the World Scientific Meeting on the Biology of Sardines and Related Species*, III—Stock and Area Papers, **4**, pp. 793–805, F.A.O., Rome.

MUŽINIČ, RADOSNA (1960b) Annual changes in the size of sardine and in the yield of the fishery in the central part of the Eastern Adriatic, *Proceedings of the World Scientific Meeting on the Biology of Sardines and Related Species*, III—Experience Paper, **6**, pp. 977–81, F.A.O., Rome.

MUŽINIČ, RADOSNA (1960c) On the identification of populations of sardine in the Adriatic, *Proceedings of the World Scientific Meeting on the Biology of Sardines and Related Species*, III—Experience Paper. **7**, pp. 983–8, F.A.O., Rome.

MacGinitie, G. E. and MacGinitie, Nettie (1949) *Natural History of Marine Animals*, McGraw-Hill, New York.

MacGregor, J. S. (1957) Fecundity of the Pacific sardine (*Sardinops caerulea*), *Fishery Bull. Fish Wildl. Serv. U.S.* **57** (121), 427–49.

MacGregor, J. S. (1959) Relation between fish condition and population size in the sardine (*Sardinops caerulea*), *Fishery Bull. Fish Wildl. Serv. U.S.* **60** (166), 215–30.

MacGregor, J. S. (1964) Relation between spawning stock size and year-class size for the Pacific sardine, *Sardinops caerulea*, *Fishery Bull. Fish Wildl. Serv. U.S.* **63** (2), 477–91.

Mackintosh, N. A. (1965) *The Stocks of Whales*, Fishery News (Books) Ltd., London.

Nair, R. V. (1960) Synopsis on the biology and fishery of Indian Sardines, *Proceedings of the World Scientific Meeting on the Biology of Sardines and Related Species*, II—Species, synopses, subject synopses, pp 329–414, F.A.O. Rome.

Nawratil, O. (n.d.) The pilchard of South West Africa (*Sardinops ocellata*). Age studies and age composition of *Sardinops ocellata* in the commercial catches, 1952–1958, and a new method for the determination of the age of *Sardinops ocellata*, *Mar. Res. Laboratory*, *Invest. Rep.* **2**, South West Africa.

Norman, J. R. (1931) *A History of Fishes*, Benn, pp. 262–3.

O'Connell, C. P. (1955) The gas bladder and its relation to the inner ear in *Sardinops caerulea* and *Engraulis mordax*, *Fishery Bull. Fish Wildl. Serv.* **56** (104), 505–33.

Orton, J. H. (1920) Sea temperature, breeding and distribution in marine animals, *J. Mar. Biol. Ass. U.K.* **12** (2), 339–66.

Pappé, L. (1853) Synopsis of the Edible Fishes of the Cape of Good Hope, p. 29. L. Taats, Cape Town, 34 pp.

Plessis, C. G., du (1958) Fishing with the South African pursed lampara, *World Fish* **7** (3), 57–8.

Plessis, C. G., du (1960) Trends in the pilchard fishery of the Union of South Africa 1943–58, *Proceedings of the World Scientific Meeting on the Biology of Sardines and Related Species*, III—Stock and Area Papers, pp. 631–66, F.A.O., Rome.

Radovich, John (1962) Effects of sardine spawning stock size and environment on year-class production, *Calif. Fish Game* **48** (2), 123–40.

Ramalho, A. (1933) Fluctuation saisonnière du poids moyen de la sardine, *C.R. Soc. Biol.* T 113, 745–6, Paris.

Ramalho, A. (1935) Sur la variation mensuelle du poids moyen de la sardine, *C.R. Soc. Biol.* T 120, Lisbon.

Raymont, J. E. G. (1963) *Plankton and Productivity in the Oceans*, Pergamon Press, Oxford.

Ricker, W. E. (1954) Stock and recruitment, *J. Fish. Res. Bd. Can.* **11** (5), 559–623.

Riedel, Dietmar (1960) Sardine production off the Atlantic coast of Europe and Morocco since 1920, *Proceedings of the World Scientific Meeting on the Biology of Sardines and Related Species*, III—Stock and Area Papers, **7**, pp. 877–911.

Riley, G. A. (1941) Plankton studies IV. George's Bank, *Bull. Bing. Oceanogr. Coll.* **7** (4), 1–73.

Riley, G. A. and Bumpus, D. F. (1948) Phytoplankton–zooplankton relationships on George's Bank, *Sears Found. J. Mar. Res.* **6** (1), 33–47.

Roach, S. W., Harrison, J. S. M. and Tarr, H. L. A. (1961) Storage and transport of fish in refrigerated sea water, *Fish. Res. Bd. Can. Bull.* 126.

Ronquillo, I. A. (1960) Synopsis of biological data on Philippine sardines (*Sardinella perforata*, *S. fimbriata*, *S. sirm*, *S. longiceps*) *Proceedings of the World Scientific Meeting on the Biology of Sardines and Related Species*, II—Species synopsis, subject synopsis, pp. 453–95. F.A.O., Rome.

Rosa, Horacio (1965) Preparation of synopses on the biology of living aquatic organisms, *F.A.O. Fisheries Synopsis* 1 (Rev. 1), F.A.O., Rome.

Rosa, H. and Laevastu, T. (1960) Composition of biological and ecological characteristics of sardines and related species, *Proceedings of the World Scientific Meeting on the Biology of Sardines and Related Species*, II—Species synopses, subject synopses, pp. 521–52, F.A.O., Rome.

Rowe, John (1953) *Cornwall in the Age of the Industrial Revolution*, Liverpool University Press.

Russell, E. S. (1931) Some theoretical considerations on the "over-fishing" problem, *J. Cons. Perm. Int. Explor. Mer.* **6** (1), 3–20.

Russell, F. S. (1930) The seasonal abundance of the pelagic young of teleostean fishes in the Plymouth area, *J. Mar. Biol. Ass. U.K.* **16**, 707.

Russell, F. S. (1935) On the value of certain plankton animals as indicators of water movements in the English Channel and North Sea, *J. Mar. Biol. Ass. U.K.* **20**, 309.

Russell, F. S. (1936a) Observations on the distribution of plankton animal indicators in the mouth of the English Channel, *J. Mar. Biol. Ass. U.K.* **20**, 507–22.

Russell, F. S. (1936b) The seasonal abundance of the pelagic young of teleostean fishes in the Plymouth area, *J. Mar. Biol. Ass. U.K.* **20**, 595.

Russell F. S. (1939) Hydrographical and biological conditions in the North Sea as indicated by plankton organisms, *J. Cons. Perm. Int. Explor. Mer.* **14**, 172–90.

RUSSELL, SIR F. S. and YONGE, SIR MAURICE (1968) *Advances in Marine Biology*, **6**. See GULLAND, J. A. and CARROZ, J. E., *Management of Fishery Resources*, 1–71, Academic Press, London and New York.

SCHAEFER, M. B. (1954) Fisheries dynamics and the concept of maximum equilibrium catch, *Proc. Gulf and Caribbean Fisheries Institute*, 6th Annual Session, November 1953, 53–64.

SCHAEFER, M. B., SETTE, O. E. and MARR, J. C. (1951) Growth of the Pacific coast pilchard fishery, *Res. Rep. U.S. Fish Wildl. Serv.* 29.

SCHMIDT, P. G. (1959) The puretic power block and its effect on modern purse seining, in KRISTJONSSON, HILMAR (Ed.), *Modern Fishing Gear of the World*, pp. 400–13, Fishing News (Books), London.

SCOFIELD, E. C. (1934) Early life history of the California sardine (*Sardinops caerulea*), with special reference to distribution of eggs and larvae, *Fish Bull. Calif.* **41,** 48 pp.

SCOFIELD, N. B. (1931) Report of the Bureau of Commercial Fisheries, *Bienn. Rep. St. Calif. Div. Fish Game* **31,** 1928–30, 106–28.

SCOFIELD, N. B. (1932) Report of the Bureau of Commercial Fisheries, *Bienn. Rep. St. Calif. Div. Fish Game* **32,** 1930–32 64–79.

SCOFIELD, N. B. (1934) Report of the Bureau of Commercial Fisheries, *Bienn. Rep. St. Calif. Div. Fish Game* **33,** 1932–34, 47–60.

SCOFIELD, W. L. (1938) Sardine oil and our troubled waters, *Calif. Fish Game* **24** (3), 210–23.

SCOTT, A. D. (1964) Food and the world fisheries, in CLAWSON, MARION (Ed.), *Natural Resources and International Development*, Johns Hopkins Press, Baltimore.

SELDON, ARTHUR and PENNANCE, F. G. (1965) *Everyman's Dictionary of Economics*, Dent, London.

SETTE, O. E. (1961) Problems in fish population fluctuations, *Rep. Calif. Coop. Ocean. Fish. Invest.* **VIII,**21–24.

SIDAWAY, E. P. (n.d.) Certain aspects of the German herring canning industry, *British Intelligence Objectives Sub-Committee, Final Report No.* 1071, Item 22, H.M.S.O., London.

SIMPSON, A. C. (1956) The pelagic phase, in GRAHAM, MICHAEL, *Sea Fisheries: Their Investigation in the United Kingdom*, pp. 207–50, Arnold, London.

SOUTH AFRICAN SHIPPING NEWS and FISHING INDUSTRY REVIEW (1960) **XV** (6), 57. Thomson Newspapers, Cape Town.

—— (1961a) **XVI** (6), 61.

—— (1961b) **XVI** (8), 51.

—— (1961c) **XVI** (9), 55.

—— (1962a) **XVII** (1), 57.

—— (1962b) **XVII** (1), 73.

—— (1964a) **XIX** (4), 91.

—— (1964b) **XIX** (7), 87.

—— (1964c) **XIX** (7), 93.

—— (1964d) **XIX** (7), 101.

—— (1964e) **XIX** (10), 79.

—— (1965a) **XX** (4), 81.

—— (1965b) **XX** (7), 81.

SOUTH AFRICAN SHIPPING NEWS AND FISHING INDUSTRY REVIEW (1965c) **XX** (7), 93. Thomson Newspapers, Cape Town.

—— (1966a) **XXI** (4), 77.

—— (1966b) **XXI** (4), 91.

—— (1966c) **XXI** (5), 89.

—— (1966d) **XXI** (6), 77.

—— (1966e) **XXI** (6), 85.

—— (1966f) **XXI** (8), 75.

—— (1966g) **XXI** (9), 119.

—— (1966h) **XXI** (10), 77.

—— (1966i) **XXI** (9), 109.

—— (1967a) **XXII** (4), 101.

—— (1967b) **XXII** (4), 103.

—— (1967c) **XXII** (5), 77.

—— (1967d) **XXII** (6), 73.

—— (1967e) **XXII** (7), 81.

—— (1967f) **XXII** (9), 75.

—— (1967g) **XXII** (10), 53.

SOUTH AFRICAN SHIPPING NEWS AND FISHING INDUSTRY REVIEW (1967h) **XXII** (11), 101, 103. Thomson Newspapers, Cape Town.

—— (1967i) **XXII** (4), 97.

—— (1968a) **XXIII** (1), 3.

—— (1968b) **XXIII** (1), 105.
—— (1968c) **XXIII** (1), 119.
—— (1968d) **XXIII** (4), 79.
SOUTHWARD, A. J. (1962) The distribution of some plankton animals in the English Channel and approaches, II. Surveys with the Gulf III high-speed sampler, 1958–60, *J. Mar. Biol. Ass. U.K.* **42** (2), 275–375.
SOUTHWARD, A. J. (1963) The distribution of some plankton animals in the English Channel and approaches, III. Theories about long-term biological changes, including fish, *J. Mar. Biol. Ass. U.K.* **43** (1), 1–30.
SOERJODINOTO, R. (1960) Synopsis of biological data on lemuru *Clupea (Harengula) longiceps, Proceedings of the World Scientific Meeting on the Biology of Sardines and Related Species,* II—Species synopses, subject synopses, pp. 318–28, F.A.O., Rome.
STANDER, G. H. (1964) The pilchard of South West Africa. The Benguela current off South West Africa, *Mar. Res. Laboratory, Invest. Rep.* **12,** South West Africa.
STANDER, G. H. (1965) Pilchard fishery trends 1950–65, *S. Afr. Shipp. News and Fish. Ind. Rev.* **XX** (7), 70.
STANDER, G. H. and LE ROUX, P. J. (1968) Notes on the fluctuations of the commercial catch of the South African pilchard (*Sardinops ocellata*) 1950–65, *Div. Fish. Invest. Rep.* **65,** Pretoria.
STEINDACHNER, FRANZ (1879) Ueber einige neue und seltene Fisch-Arten aus den K.K. Zoologischen Messen zulvien, Stuggart und Warschau, *Akad. Wiss. Wien. Deukschr.* **41,** 12–13.
STEEMANN NIELSEN, E. (1937) On the relation between the quantities of phytoplanton and zooplankton in the sea, *J. Cons. Int. Explor. Mer.* **12,** 147–54.
STEVEN, G. A. (1948) Contributions to the biology of the mackerel, *Scomber scombrus* L. Mackerel migrations in the English Channel and Celtic Sea, *J. Mar. Biol. Ass. U.K.* **XXVII,** 517–39.
STODDART, D. R. (1965) Geography and the ecological approach. The ecosystem as a geographic principle and method, *Geography* **50** (3), No. 228, 242–51.
STOLTING, W. H. and MURRAY, A. T. (1958) Research on costs and earnings in the commercial fishing industry in the United States, in *Report of the Technical Meeting on Costs and Earnings of the Fishing Enterprises,* F.A.O., Rome.
STOOPS, W. H. (1953) The South African fishing industry, *S. Afr. J. Econ.* **21** (3), 242–50.
SVERDRUP, H. U., JOHNSON, M. W. and FLEMING, R. (1942) *The Oceans. Their Physics, Chemistry and General Biology,* Prentice-Hall, New York.
SVETOVIDOV, A. N. (1952) *Fauna U.S.S.R. Fishes, 2 (1) Clupeids,* Moscow, 331 pp.
SWITHINBANK, H. and BULLEN, G. E. (1914) The scientific and economic aspects of the Cornish pilchard fishery 2. The plankton of the inshore waters in 1913 in relation to the fishery, *Mera Publications* 2, St. Albans.
TAYLOR, H. F. (1955) Food productivity of the ocean, *Oceanus* **IV** (1), 12–17.
TAYLOR, R. A. (1960) *The Economics of White Fish Distribution in Great Britain,* Duckworth, London.
TEMMINCK, C. J. and SCHLEGEL, HERMAN (1850) Pisces, in *Fauna Japonica, sine descriptio animalium quae in itinere per Japonicum suscepto annis 1823–1830 collegit, etc. by von Siebold,* Lugdini, Batavorum, 237.
THOMPSON, W. F. (1924) Report of the State Fisheries Laboratory, *Bienn. Rep. St. Calif. Fish Game Commn.* **28,** 1922–4, 56–57.
THOMPSON, W. F. (1926) The California sardine and the study of the available supply, *Fish. Bull. Calif.* **11,** 7–66.
THOMPSON, W. F. (1929) The regulation of the halibut fisheries of the Pacific coast of North America, *J. Cons. Perm. Int. Explor. Mer.* **4** (2), 145–61.
TOKYO (RESEARCH LABORATORY) (1960) Synopsis on the biology of *Sardinops melanosticta, Proceedings of the World Scientific Meeting on the Biology of Sardines and Related Species,* II—Species synopses, subject synopses, pp. 213–44, F.A.O., Rome.
TREBETT, DAVID (Ed.) (1967) *The South African Fishing Industry Handbook and Buyers' Guide, 1966/67,* Thomson Newspapers, Cape Town.
TYSSER, H. F. (Ed.) (1960) *Fisheries Year-Book and Directory 1960,* British-Continental Trade Press Ltd., London.
VON BRANDT, ANDRES (1964) Fish Catching Methods of the World, pp. 144–5, Fishing News (Books) Limited, 191 pp.
WALBAUM, I. I. (1792) Petri Artedi sueci genera piscium .... Eneudata et aucta a Iohanne Iulio Walbaum. *Grypeswaldiae* **3,** 38.
WHEELER, ALWYNNE (1964) *Sardina pilchardus* (Walbaum, 1792): Proposed preservation as the name for the European sardine (Pisces), *Bull. Zool. Nomencl.* **21** (5), 360–2.
WILCOCKS, J. C. (1883) The history and statistics of the pilchard fishery in England, in HERBERT, DAVID, *Fish and Fisheries,* pp. 290–303. Blackwood, Edinburgh.
WOLF, R. S. (1961a) Graphic presentation of Pacific sardine age composition data, *Spec. Sci. Rep. U.S. Fish Wildl. Serv.* **384,** 1–34.
WOLF, R. S. (1961b) Age composition of the Pacific sardine, 1932–1960, *Res. Rep. U.S. Fish Wildl. Serv.* **53,** 1–36.
WOLF, R. S. (1964) Observations on spawning Pacific sardines, *Calif. Fish Game* **50** (1), 53–57.
YOUNG, J. Z. (1962) *The Life of Vertebrates,* Oxford University Press.

INDEX

(References in italic indicate that there is an illustration on that page)